Understanding Purpose

North American Kant Society Studies in Philosophy

General Editor
Robert B. Louden
University of Southern Maine

Editorial Board

Sharon Anderson-Gold
Richard Aquila
Andrew Brook
Michael Friedman

UNDERSTANDING PURPOSE

KANT AND THE PHILOSOPHY OF BIOLOGY

Edited by Philippe Huneman

Volume 8
North American Kant Society
Studies in Philosophy

UNIVERSITY OF ROCHESTER PRESS

First published 2007

University of Rochester Press
668 Mt. Hope Avenue, Rochester, NY 14620,
www.urpress.com
and Boydell & Brewer Limited
PO Box 9, Woodbridge, Suffolk IP12 3DF, U
www.boydellandbrewer.com

ISBN-13: 978–1–58046–265–5
ISBN-10: 1–58046–265–0

Library of Congress Cataloging-in-Publication Data

Understanding purpose : Kant and the philosophy of biology / edited by Philippe Huneman.
    p. cm. — (North American Kant Society studies in philosophy, ISSN 1071–9989 ; v. 8)
    Includes bibliographical references and index.
    ISBN-13: 978-1-58046-265-5 (hardcover : alk. paper)
    ISBN-10: 1-58046-265-0
    1. Biology—Philosophy. 2. Kant, Immanuel, 1724–1804. I. Huneman, Philippe.
QH331.U53 2007
570—dc22

2007003199

A catalogue record for this title is available from the British Library.

This publication is printed on acid-free paper.
Printed in the United States of America.

CONTENTS

Acknowledgments   vii

Editor's Notes   ix

Introduction   1

1   Pre-Kantian Revival of Epigenesis: Caspar Friedrich Wolff's
    De formatione intestinorum (1768–69)   37
    Jean-Claude Dupont

2   Kant's Persistent Ambivalence toward Epigenesis, 1764–90   51
    John H. Zammito

3   Reflexive Judgment and Wolffian Embryology: Kant's Shift
    between the First and the Third Critiques   75
    Philippe Huneman

4   Kant's Explanatory Natural History: Generation and
    Classification of Organisms in Kant's Natural Philosophy   101
    Mark Fisher

5   Succession of Functions and Classifications in Post-Kantian
    Naturphilosophie around 1800   123
    Stéphane Schmitt

6   Goethe's Use of Kant in the Erotics of Nature   137
    Robert J. Richards

7   Kant and British Bioscience   149
    Phillip R. Sloan

Bibliography   171

List of Contributors   185

Index   187

ACKNOWLEDGMENTS

The origin of this volume is a pair of symposia that I organized in 2003, the first one at the International Society for History and Philosophy and Social Sciences of Biology meeting in Vienna (July) and the second one at the International Society for Eighteenth Century Studies at UCLA (August). First of all, I wish to thank the speakers of those sessions for having accepted to contribute to this work, therefore making such a volume possible, and for all the exchanges we had on several occasions about those subjects. I wish also to thank particularly the four American authors for having accepted to revise the English language of the papers written by the French scholars.

I am glad to thank the North American Kant Society, which agreed to publish this book in this collection. Above all, I am grateful to Robert Louden, who also helped me during the whole process, and Eric Watkins, who initially suggested to me to submit the book to the NAKS Publication Series.

Among all the people to whom I am indebted for having exchanged ideas about those topics during the three last years, I cannot but mention Christophe Bouton, Daniel Dumouchel, Peter McLaughlin, Claude Piché, and Joan Steigerwald. I am also hugely indebted to Gérard Lebrun, who has been for me a master and introduced me to the philosophical sense of the history of philosophy. Unfortunately, he died in 2000; I consider that my own work in this field would not have been possible without him and I hope that this work will contribute to his memory.

Finally, I warmly thank Bob Richards, who invited me to teach on Kant and biology at the Committee for Conceptual and Historical Studies on Science at the University of Chicago in 2005. I greatly benefited from all the discussions that took place there, both in my class and in a seminar we had on the third *Critique*. I can't cite everybody, but I'm sure I must acknowledge Julia Hylard Bruno, Chris Di Teresi, Trevor Pearce, Bill Sterner, Kathryn Tabb, Thomas Teufel, and Ken Westphal.

As usual, all mistakes in my chapters are mine.

P. H., Paris

EDITOR'S NOTES

In this volume, the *Critique of Judgment* is cited as *KU*. References to Kant's works are given in the *Akademie Ausgabe* edition. When a specific translation is cited in a chapter, the authors will give the particular reference. References to the *Critique of Pure Reason*, as usual, are given by citing the original page numbers from editions A and B (e.g., A546/B578). By default, the translation used here is the *Critique of Pure Reason*, translated and edited by Paul Guyer and Allen Wood (Cambridge: Cambridge University Press, 1998).

The most cited and relevant works of Immanuel Kant concerning the topic of purposiveness and biology considered in this volume are the following.

Kant, Immanuel (1755). *Allgemeine Naturgeschichte und Theorie des Himmels*. Akademie-Ausgabe, vol. 1: 217–368.

———(1763). *Der einzig mögliche Beweisgrund zu einer Demonstration des Daseins Gottes*. Akademie-Ausgabe, vol. 2: 63–164.

———(1775). "Von den verschiedenen Racen der Menschen." Akademie-Ausgabe, vol. 2: 427–44.

———(1785a). "Recension von J. G. Herders *Ideen zur Philosophie der Geschichte der Menschheit. Theil 1. 2.*" Akademie-Ausgabe, vol. 8: 43–66.

———(1785b). "Bestimmung des Begriffs einer Menschenrace." Akademie-Ausgabe, vol. 8: 89–106.

———(1788). "Über den Gebrauch teleologischer Principien in der Philosophie." Akademie-Ausgabe, vol. 8, 157–84. Translation by J. Mikkelsen. In Robert Bernasconi and T. L. Lott (eds.), *The Idea of Race*. Indianapolis: Hackett, 2000: 8–22.

———(1790). *Kritik der Urteilskraft*. Akademie-Ausgabe, vol. 5: 165–486.

INTRODUCTION

KANT AND BIOLOGY? A QUICK SURVEY

PHILIPPE HUNEMAN

Kant's *Critique of Judgment* is known as a major metaphysical achievement concerning the concept of *purpose*. Under the term *Naturzweck*, Kant tries to legitimize the use of purposive categories in the life sciences without any theological commitment, notwithstanding the whole philosophical tradition from stoicism to natural theology. Between the Spinozist or Humean critique of finality as an illusory concept—conceived either from a Cartesian-rationalist or an empiricist point of view—and the Leibnizian-Wolffian assumption of finality as the link between natural history and theology, Kant recognized an epistemological value of the concept of finality while denying its major ontological implications concerning some creative intention at the source of the universe and of living beings. The natural purposes, *Naturzwecke*, are organisms, or organized beings, *Organisierte Wesen*.[1] This main thesis of the *Critique of Judgment* meant that living creatures could—indeed, must—be conceived as purposive entities, that is, as possible solely according to a concept, but that this concept need not to be instantiated in reality; it merely has to be presupposed for us as a principle of knowledge.[2] Kant called this supposition a "regulative use" of the concept of finality. Hence, the epistemological requirement of a specificity for the knowledge of living beings is satisfied, while at the same time, no commitment is made to something outside the reach of natural science and its methods. Therefore we can ask: To what extent did Kant really *naturalize* purpose?

This question is as much an issue for the history of philosophy as for the history of science. Kant's metaphysical elucidation of purpose stemmed, of course, from his previous considerations about the possibility of knowledge, realism concerning an order of nature, limits of reason, and so on. As such, however, it concerns some major advances in the life sciences, such as in descriptive embryology or comparative anatomy. This is attested to by Kant's various quotations, references, and discussions of scientists in those domains; Blumenbach[3] and the major anthropologists Plattner and Hufeland;[4] anatomists such as Soemmering;[5] naturalists such as Buffon, Maupertuis, and Erxleben;[6] and the four great men of eighteenth-century physiology: Boerhaave, Stahl, Hoffmann, and Haller.[7]

In some sense not only the third *Critique*, but also Kant's various papers on biological issues, can be seen as elements in a research program for the life sciences, given that their central concept, *Naturzweck*, is intended to fulfill requisites of both epistemological autonomy and naturalistic orthodoxy for those sciences. Hence the research program is based on two correlative claims: there is a teleological principle as heuristical framework for the mechanistic analyses of causal processes in organisms,[8] and there is an irreducibility of the source of this teleology—called "original organization" by Kant[9]—to mechanistic devices. Our question becomes then: to what extent does this program fit the sciences of its time?

We know that the word "biology" was forged in the first years of the nineteenth century, by Lamarck, Burdach, and Treviranus.[10] This novelty is an indication of something new at stake in the conceptual domain. It is noteworthy that, at the same time, Kant proposed an "analytics" for biological sciences within his wide metaphysical questioning. This is also certainly full of valuable insights concerning the major problems raised by the foundation of what emerged as a science of life in this period.

In his influential works, Timothy Lenoir contends that further developments in German biology can be addressed along Kantian lines. In fact, he tries to isolate a research program, named "teleomechanism," that is initiated by Kant and carried out by biologists in the Göttingen school, from Blumenbach to Johannes Müller through Reil and Von Baer. This program could account for the first monument of embryology—Von Baer's *Entwicklungsgeschichte der Thiere* (1828)—Müller's widely influential vitalism, the birth of cellular theory in reaction toward it, the more aggressive reaction of Du-Bois Reymond and Liebig's mechanistic program in physiology, and Von Baer's late reluctance to accept Darwinism. Under the other name of "vital materialism," such a research program, settled at its origin by both Kant and Blumenbach, could regulate the use and set the limits of the so-called mechanistic explanation in embryology and physiology, so that the specificity of living phenomena could be respected. Hence there occurred an acceptance of epigenesis in embryology, the elaboration of a theory of "types" as various ends of development, and a theory of logical transformation of these types that is both close to transformism (since it is quite historical)[11] and reluctant to Darwinism (since the types are immutable ideas).

Of course, this account has been repeatedly challenged in the last twenty years. Historians such as Kenneth Caneva, Phillip Sloan, Robert Richards, and James Larson, while working on these topics, contest Lenoir's reconstruction. In these debates, numerous issues included in this period of the development of biology have become increasingly clear. In particular, it now seems that the relationship between Kant's philosophical project and the projects both of contemporary biologists and of the *Naturphilosophie* that takes place a bit later have been misconceived by Lenoir. Recently, Robert J. Richards's reconstruction of this biology in his *Romantic Conception of Life*, more comprehensive because it considered both Goethe's work and *Naturphilosophie*, tried to give a more precise picture of the theories and the debates in which Kant's naturalized purpose took part. This means that the time has come for a critical exposition of some of these issues and of some of the results of this twenty-year-long series of research in the history of science and philosophy. One of the results, of course, is that the line we are prompted to draw between science and philosophy appears quite blurred, or more complicated than we thought, since people at those times would spontaneously elaborate their explanatory strategy within both science and philosophy.

Above all, it appears that several fields of discourse were connected by the so-called teleomechanist research program while they were in elaboration:

— Embryology, of course, from Caspar Wolff[12] to Von Baer,[13] through the major works of Blumenbach.
— Medicine and the anthropology that was called "natural history of man,"[14] whose practitioners were mostly physicians in Germany and which was linked to the idea of *Populärphilosophie*, a metaphysical project once shared by Kant before his critical turn.
— Morphology, represented primarily by Goethe and then Geoffroy Saint Hilaire in France and his German partisans (Meckel, Carus, later Gegenbaur).[15]
— *Naturphilosophie*, whose relationship to what could be called a Kantian research program in biology is an important part of the questions raised by studies of this period.

This sets the stage for an explication of some major features of the epistemological configuration involving Kantianism and early nineteenth-century German biology. As this volume is a kind of "review" of the perspective in this field gained by historiographical research in the last two decades, I first provide a quick survey of these features as they have been discovered and discussed in the works of philosophers and historians of science, mainly from Lenoir's *Strategy of Life* to Richard's *Romantic Conception of Life*.

First, I consider some issues in Kant's thought and the development of his philosophy, regarding the history of sciences before and around him. Then, I survey some issues regarding the Kantian legacy in the philosophy of biology, and the interpretation of Kant by his immediate followers, namely, a set of questions focusing on the possibility of speaking of a "Kantian tradition in biology."

1. Issues in Kant's Thinking about Biology

1.1. Systematicity of Knowledge and Biology

As stated in the introduction to the *Critique of Judgment*, one of Kant's main concerns with biology regards the possibility of classifying species of organisms, a requisite stemming from the program of natural history.[16] The only physiological examples in the first *Critique* are therefore to be found in the "Appendix to the Transcendental Dialectic,"[17] which is the part of the *Critique of Pure Reason* devoted to an elucidation of the order of nature, given that, according to the Transcendantal Analytic, the most universal laws of nature are no longer thought to be grounded in a divine architect but, rather, in the mere human understanding.

Because of the necessity to classify, and because this biologically primary feature called "heredity" or "conservation of form" requires classification and separation of species, a philosophical perspective on the life sciences should be situated within the general problem of the order of nature. This metaphysical problem, transferred to a transcendental framework—that is, a question about conditions of knowledge rather than kinds of being—becomes the epistemological problem of the *systematicity of knowledge*.[18]

Systematicity poses the following question: To what extent are we allowed to divide nature between kinds and species, and to group individuals into species and genera? Kant remarks that unless the presupposition of the legitimacy of this activity obtains, we are by no means capable of forging empirical concepts, which are ways to group together the manifold met in experience and to name it.[19] The "Appendix" claims that this presupposition not only is formal and logical, methodological and subjective, but concerns the objectivity of the phenomena. The presupposition is, thus, transcendental, since in order to elaborate or choose any methodology, we have to assume that the world is set up in such a way as to be investigated.[20] Regardless of the differences, the introduction to the *Critique of Judgment* continues to maintain the objectivity and non-empirical status of the ability to divide nature into coalescent species and under grouped genera, for example, to order it.[21] Of course, inheritance and the generation process proper to the living realm are an important part of the requisite of considering the genera as realities.

The major texts dealing with these problems are the "Appendix" of the Dialectic of the first *Critique*, and the two introductions to the *Critique of Judgment*. One puzzle is why Kant treated it in the former text from the point of view of *reason*, and in the latter, from the point of view of the *faculty of judgment*. Further, finality and systematicity were rather synonymous in the first text, while in the *Critique of Judgment* this systematicity becomes just a kind of finality—namely, objective formal purposiveness. From within the Kantian historiography, this problem has long been debated. Bennett in

Kant's Dialectic and Walsh in *Kant's Criticism of Metaphysics* stress the inco-
herence of Kant's doctrine of systematicity in the Dialectic. While conflating
the two versions, Kitcher tries an interpretation more faithful to Kant's by
unifying finality and systematicity in his texts.[22] In "Causality, Causal Laws
and Scientific Theory in the Philosophy of Kant," Buchdahl uses both texts
indifferently. Guyer in "Reason and Reflective Judgment: Kant on the Significance of Systematicity," and Lebrun in *Kant et la fin de la métaphysique* are
more concerned by the transition.

However, the texts from the *Critique of Judgment* are slightly more concerned with the status of biology. Kant mentions Linnaeus's *Systema naturae*,
and he questions primarily the "particular analogy"[23] between the internal
constitution of members of the same species—that is, the objectivity of
grouping individuals having the same appearance (for us) within the same
species.[24] The 1788 paper, "On the Use of Teleological Principles," had been
written between these two texts as a reply to Georg Forster's paper, "Noch
etwas über die Menschenrassen."[25] It relies on a new idea of teleology, and
explicitly takes into account Kant's elaborations on the biological problems
of generation and adaptation (published in the two essays of 1775 and 1777),
with the so-called theory of *Keime* and *Anlage*.[26] Perhaps recent interpret-
ers, who focus on the relationships of these texts with natural history, find
a greater consistency in both the Dialectic's doctrine and Kant's general
progression on this problem. Among them, McLaughlin in *Kant's Critique
of Teleology* equates the two accounts, assuming that Kant did no more than
change his vocabulary.

With this important issue of systematicity, order of nature, and the place of
natural history in Kant's thinking, we are about to grasp why the life sciences
play a puzzling role in the architecture of science according to transcendental philosophy. Before turning to address that role, I summarize Kant's positions on biology.

1.2. Kant's Theoretical View of Biology

For the purpose of this preliminary exposition of the issues of this volume, I
stress a few theses that can be seen as the core of Kant's conception of biology, taken without regard to chronology.

a. The science of nature has to explain everything according to the
mechanical principle. This means that the presence and behavior of parts
account for the nature and behavior of wholes. However, some natural entities resist such an explanation. These entities display a specific relationship
between wholes and parts, in which the parts seem to presuppose the whole
in order to be accounted for. An example of this would be the regeneration
of a tree, or the embryogenesis of a chick, in which the fact that a particular

part becomes an eye and not a leg seemingly requires the previous presence of the whole, albeit not like technical production, where the whole actually exists as an idea in someone's mind, an idea that is efficacious through human technology.[27] Such entities appear to be self-organizing. They are what we call "organized beings" or organisms.

b. This presupposition of the whole is of a different status from the relationship of parts to whole in mechanical causality, where the parts determine the whole. Judging in this manner means making *teleological judgments*, or judgments of finality, since purposiveness means precisely this causal relation where the representation of the effect is the cause of the cause, or more precisely, the concept of the object lies at the ground of the possibility of the object. This relation is mainly ideal, since an idea of the whole is presupposed by us, and not judged as actually effective in nature. Therefore, such a new causality can stand beside mechanical causality and does not exclude it. Causal explanations of this type can stand together with mechanical explanations. The teleological principle here is not a principle of nature—such as the "principle of causation" that Kant contends he has founded on the possibility of experience in the first *Critique*, and on which the principle of mechanism somehow hangs. It is merely a principle of knowledge (*Erkenntnisgrund*)[28] used to help us to forge explanations of those peculiar entities not explanatorily exhausted by the fundamental principle required by science (namely, the principle of causation). Biological knowledge always combines these two registers of explanation. Embryology, as the study of these whole-directed processes, is here paradigmatic.

c. Kant holds a theory of generation—mainly expressed in the essays on races—according to which the fact that the offspring of two chicken is a chick should be explained by the development of some germs or dispositions embedded in the embryo, and selectively activated by the environment through a process of interaction that is teleologically guided (since among all dispositions, those that are best fit for this organism in his particular environment will be selected). These germs and dispositions define the identity of the species. Since the environment cannot cause them, they are not likely to vary on a wide range, and transformation of species is precluded. What this embryo will be, however, is not contained in it from the beginning (either the conception or before the conception), because it is the product of an interactive process with its particular environment.[29]

d. The set of germs and dispositions, unlike in classical preformationism, is not any visible or determined form. It defines an "original organization," which means that this whole is presupposed by the teleological judgment as directing the biological processes in an organism, processes that happen therefore in accordance with the principle of mechanism, and to this extent are likely to be explained. How this sort of agreement between kinds of explanation is possible and efficient is a difficult issue investigated in the Dialectics and the Methodology of the third *Critique*. Since this "original organization" is tran-scendentally presupposed before any account of embryological or physiologi-cal process taking place in an organism, it cannot itself be the product of any physical process. Hence, the organization of organized beings, taken both as a specific state of matter and as a disposition to organize itself in such and such a way, is not deducible from the essential properties of mere physical matter. We should notice that this boundary between life and matter is stated by Kant on the ground of epistemological rather than ontological considerations. In the context of his critical philosophy, nothing would provide us with meaningful statements concerning the ontological status of those domains.

In this reconstruction, I have passed over the details of Kant's position. It is intended merely as a basis for the reader entering the field, in order to approach the issues of interpretation and commentary here addressed.

1.3. Questions about Natural History

1.3.1. Meanings of Natural History. The first part of the eighteenth cen-tury saw the culmination of the program of naming and classifying natural entities, which is called "natural history"—essentially with the achieve-ments of Tournefort, Jussieu, Adanson, and Linnaeus.[30] It is worth noting that this ideal of classification extended even to diseases: Linnaeus is also the author of a *Genera morborum* (1763), while Boissier de Sauvages in his influencing *Nosologie méthodique* (1737) refers in the title to "la méthode des botanistes." The idea of a systema naturae faced two problems: one is methodological, the other is ontological. Methodologically, are the crite-ria used by the botanists supposed to be as important as their rank is high in the substructures of taxa? If this is the case, then we have a natural classification, called méthode, as sought by Ray or Adanson; if it is not the case, we have an artificial classification, called système, like Linnae-us's, which is grounded on the sexual characters of plants. The ontological problem appears if we proceed on the basis of those questions: What is a genus? A species? Are those categories only descriptive ones, or are they ontologically meaningful?[31] Here, we meet the debate between Buffon and Linnaeus. Buffon agressively challenged Linnaeus's system, in the name of both the continuity of nature and the unreality of taxa higher than spe-cies.[32] Like others of his generation, Kant was deeply influenced by Buf-fon's *Natural History*, one of the great editorial successes in the European academic world. After 1770, German authors esteemed particularly Buf-fon's definition of species by the reproductive viability of hybrids.[33] Later, Friedrich Girtanner wrote a natural history—*Über das kantische Prinzip für die Naturgeschichte* (1796)—deeply inspired by Kant and Buffon.[34]

This is why Buffon's criterion for species is quoted at the beginning of the first essay on human races. Further, the idea of natural history is elaborated by Kant in many places, primarily in the context of anthropology and geography.[35] He draws an important distinction between natural history as artificial description—à la Linnaeus—named *Naturbeschreibung*, and a kind of natural history that would take into account the reality of its taxa, hence the filiations among the individuals within the species, so a natural history à la Buffon, which hence would be named—given the hereditary criterion of species—"a history of nature," *Geschichte der Natur, Naturgeschichte*.[36] One may prima facie notice that a tension emerges between this program, and the idea of an order of nature that mentions no history.[37] In those two cases, a realistic commitment of science is advocated by Kant. So an important issue raised by these ideas is the connection and reconciliation of the concepts involved in a history of nature, and the transcendental requisite of an order of nature stemming from the ideas of reason. Here lies what is at stake in the later German *Naturphilosophie* and its split with Kantianism.

1.3.2. Anthropology. The issues on species are related to the definition of a human race and thereby, to anthropology. The late Eighteenth century was an important laboratory for the creation of an anthropological research program, embedded in natural history and medicine, which could be called (according to Cabanis or Virey)[38] a "natural history of man." Recent studies demonstrate the fecundity of this period in Europe concerning many domains related to a scientific approach to the study of mankind.[39] In Germany, there arose a peculiar kind of anthropological agenda. The demarcation between animal and man and the typology of races were leitmotives of this discourse.[40] Plattner's *Menschenkunde für Ärtze und Weltweise* (1772) was a success in this field and was clearly a ground-breaking work that inspired both Herder and Blumenbach.[41] Blumenbach began his career with a dissertation on races (*De generis humani varietate nativa*, 1775).[42] In some ways, Kant's treatise on human race was an answer to Plattner. Before him, Krüger and Unzer wrote widely read treatises on this medical-philosophical matter.[43] These considerations comprise the framework for Kant's investigations in *Anthropologie im prakticher Hinsicht* (1796). The relationship of this anthropological part of philosophy to Kant's transcendental philosophy is an important and internal question in his system;[44] nevertheless, the materials discussed there—medical theories about the natural history of man—and also found and developed in the volumes of lectures on anthropology (Ak XIV–XV) were developed in such an anthropological context.

Leaving aside medicine and practical advices on the good life, which were also part of this anthropology, the quarrel concerning the difference between races appeared central to this topic. Are they subspecies of a single species or various species? There was speculation concerning whether there was one or several original human couples. Of course, equality or inequality, slavery and other important political commitments were at stake in this issue. Polygenism, advocated by Voltaire and later in 1774 by Henry Home (Lord Kames) in *Six Sketches on the History of Man*, had been challenged by Buffon's monogenism.[45] Great philosophers and writers of the century (e.g., Hume) felt compelled to take sides in this quarrel. Kant's first paper on races was of course intended to establish the unity (or singularity) of the human species. This debate stimulated Blumenbach, Forster, and others.

It is important to be reminded that Kant's position has to be read in contrast with those of other writers in this large debate. This was, perhaps, a reason for publishing the essay on the different races of men in an anthology of *Popularphilosophie* by Engel in 1777. Perhaps Kant also felt that this subject was important enough for his contemporaries that he felt compelled to reply to Forster's misunderstanding of his essay, and then to write the *Gebrauch* essay ("On the Use of the Principles of Teleology") in 1788.[46]

Briefly put, none of Kant's biological concerns should be separated from this important debate on races, which was at the heart of the construction of a German anthropology at the end of the century.[47]

1.4. Regulative and Constitutive Principles

It is important to emphasize that all these considerations are presented by Kant as part of an enquiry about regulative principles of science. The principles of species and genera in the "Appendix," as much as presuppositions of the purposiveness of nature linked to the faculty of judgment in the *Critique of Judgment*, and the principle of natural purposes as organisms,[48] are seen by Kant as regulative principles. Accordingly, they are opposed to constitutive principles, that is, principles that are constitutive of phenomena as such, and not merely of the rational knowledge of phenomena—given that, in the transcendental perspective, the lawlikeness of the phenomena is constituted by the principles of our understanding.[49] However, this basic conceptual distinction in Kantian thought has been quite puzzling for commentators. Things get worse if we remember that Kant also uses this distinction to present the difference between two kinds of principles of the understanding, the mathematical ones and the dynamical ones.[50] This means that this important distinction must always be grasped in its proper context.

Contrasting visions of Kant's philosophy of biology have been built from the meaning given by the commentators to regulative principles. If the word "regulative" means that those principles are only to help us in ordering the phenomena, they are a kind of heuristic. This is a weak reading of Kant's philosophy of biology, very close to Vaihinger's view in *Philosophie des als ob*

(1906): we must proceed as if living nature were purposive, but we know that it is not. Notice that this widely diffused reading is well suited to some kinds of reductionism that tolerate teleological notions only as heuristical tools. Such an interpretation, however, becomes problematic when we understand that the whole "Appendix to the Transcendental Dialectic" tries to prove that principles of homogeneity of nature, of specification, and so on, are *not* merely methodologically requisite, but have an objective bearing.[51]

Another reading of those concepts insists, then, on the fact that the necessity of regulative principles is not of methodological character—as a shorthand description of phenomena or a kind of principle of economy of thought—but stems from the finite character of our faculty of knowledge and concerns the *ability* of this faculty to reach knowledge, and not solely its best or simplest way to reach it. This reading could seem convincing, but is quickly threatened by the temptation of identifying the two kinds of principles, the constitutive and the regulative, under the pretext that both express some conditions of knowledge and experience for a finite rational being.[52] This inclination could lead us directly to Hegel and German idealism—and perhaps Kant followed it in the *Opus postumum* (see below).

Moreover, any attempt of interpreting Kant prima facie presupposes coherence in his concepts and continuity in the development of his thought. But this anticipation is not an axiom, and can be either corroborated or falsified by reading the texts. It may be that with respect to the constitutive/regulative issue, such a presupposition will have to be qualified.[53]

One's interpretation of the antinomies depends on one's choice among these options. In the "Critique of Teleological Judgment," we have an antinomy between mechanism and teleology, which is solved through an appeal to the difference between regulative and constitutive principles. It is not clear, however, whether the principle of mechanism has to be judged as a constitutive principle (i.e., as another version of the so-called principle of causality stated in the Second Analogy), or whether it is a merely regulative principle, even if the principle of causality, on which it is grounded, is a constitutive principle.[54] The solution of the antinomy of judgment is thus opaque, as is noted by McLaughlin.[55] Are the two maxims—of teleology and of mechanism—related to reflexive judgment, or is it only the case of the teleological principle, as maintained by Heinrich Cassirer?[56] The problem is that if the two maxims are regulative, there would be no antinomy at all (like the maxims of specification, homogeneity, and continuity of genera and species in the "Appendix to the Transcendental Dialectic").

"Antinomy" is of course a technical word of the first *Critique*. Kant exposes the antinomy of cosmology, which is linked to the *Idea of a World*. However the section entitled "Goal of the Natural Dialectic," following the Dialectic of the *Critique of Pure Reason*, does not mention any "antinomy" of teleology, yet there is a kind of reconciliation between "mechanism" and "purposiveness" as defined in this text by reference to the Idea of God.[57] Kant says that those two principles are not conflicting; they are in fact harmonious because they eventually yield the same strategy of explanation.[58] The genuine antinomy involving mechanism is the famous Third Antinomy of the Dialectic, which opposes mechanism to free causality.[59] Hence, antinomy of reason means either an opposition between mechanism as a principle of judgment and the *lack of* natural causality in a series of events (Dialectic of the first *Critique*) or between mechanism and *final* causality (Dialectic of the third *Critique*). Both are solved in the same manner by appealing to the regulative nature of the principles. Nevertheless, this does not tell us whether the *Critique of Judgment* presents an "antinomy" of the same nature as the one of the first *Critique*.[60] Deciding these hermeneutic points requires, above all, a clarification of just what "regulative" and "constitutive" mean.[61] These issues are of the utmost concern for interpreting Kant's critical philosophy. Historically, they have led to a major parting of the ways of life scientists of those times concerning the significance of Kant's legacy (see below).

As purposiveness is the regulative principle of the faculty of reflexive judgment, and this judgment operates in the aesthetic, biological, and logical spheres, one has to make clear the varied meanings of purposiveness in Kant. In fact, this regulation of judgment is not exactly the same when it bears on the form or on the content of the objects—hence the notions of "formal" and "material" purposiveness, as well as "objective" and "subjective" purposiveness. Following Marc-Wogau are Giorgio Tonnelli, George Schrader, John MacFarland, and Gérard Lebrun, among others, who have made some precise comments and distinctions on this matter.[62]

1.5. Epigenesis

1.5.1. The Debate Concerning the Generation of Organisms.

As a regulative principle, the idea of *Naturzweck* is closely connected with the idea of epigenesis and, broadly speaking, with issues about generation, given that according to this principle the parts of a natural purpose have to cause each other, that is, bring one another into being.[63] Both in the *Critique of Judgment* and in the texts on race and on teleological principles, Kant elaborates a theory of the origin of living organisms, in the context of the epigeneticism/preformationism debate.[64] In Kant's time, this debate is already nearly two centuries old. Harvey, following Descartes, held epigenesis as a mechanistic description of the generation of bodies. Preformationism then became dominant in the seventeenth century, being argued for empirically with appeal to the microscope by Leuwenhook, after having been defended on metaphysical grounds by Leibniz and Malebranche.[65]

The ideas of "teleomechanism" (Lenoir) and of a "Kantian research program in the life sciences," whether or not they are historically valuable, emphasize the importance of issues concerning heredity and generation in Kantian thinking. The relationships between Kant and Blumenbach have been a major topic in the historiographical literature.[66] The question is twofold: What does Kant's commitment to epigenesis in the third *Critique* mean, and what is the nature of this commitment? What is the relationship between his position—and the germ-theories that are its corollaries—and the strongest positions in the field, namely, Blumenbach's, Wolff's, and Haller's? This second question requires further elaboration concerning the inverse relationship between Kant's position and the fate of embryology in German biology from Blumenbach to Von Baer. (Of course, all of these questions suppose that we have reconstructed what it means to be an "epigeneticist" in the late eighteenth century.) Such a question has been the focus of Lenoir's books, but his answers demand qualifications. I sketch the major issues in the next section.

One interesting point concerns the relationship of epigeneticism and vitalism, considered as the global claim that there must be a vital force. Kant claimed in his own way the truth of a kind of epigeneticism, but refused a kind of vitalism, which he termed "hylozoism," that ascribes life to mere matter.[67] Nonetheless, Peter Hans Reill argues that in the mid-eighteenth century against mechanicism and Newtonian science a quasi-skepticism about causality allowed the rise of an "Enlightenment vitalism," stressing the irreducibility of life force.[68] The point, then, is to situate Kantianism within this trend of thought, which made a significant use of vital forces and tried to correlate them.[69]

While Lenoir insists on the continuities between Blumenbach and Kant (following Kant's and Blumenbach's own acknowledgment of their indebtness toward each other), Richards highlights the differences between them.[70] In his review of Lenoir's book, Caneva sharpens this contrast.[71] One must now recognize that Lenoir's attempt to isolate a Blumenbachian-Kantian core of a research program is rather a retrospective reconstruction, and that the divergence he wanted to highlight in the subsequent history was emerging from the very beginning between the two original instigators of this research program. Of course, one main issue involved here is the relationship between naturalistic research and metaphysical investigation such as Kant's transcendental philosophy. Like many others, the authors of the present volume agree that Kant's concern was primarily metaphysical, even while dealing with biology—which of course is not shared by Blumenbach and other members of what Lenoir in a paper called the Göttingen school in the life sciences.

One main concern for the interpretation is Kant's strange terminology to name his position because, whereas he rejects individual preformationism, he calls himself—in the context of his philosophical denial of preformation-ism—a "generic preformationist."[72] This puzzle is also the focus of chapter 4 in this volume.

1.5.2. Kant's Use of "Epigenesis" in the First *Critique*. In the second edition of the first *Critique* (B167), Kant speaks of "epigenesis" in order to charac-terize the original configuration of knowledge by the categories. This text gives birth to an important question for the interpretation of Kant's theory of knowledge. Ingensiep, Wubnig, and Zöller, after Genova,[73] wonder why Kant had to compare this activity of categories to the biological process of epigenesis, and, moreover, whether this is a merely comparison or whether it describes another kind of actually epigenetic process.[74] They try to consider what "epigenesis" could mean to Kant at the time he was writing and in rela-tion to the leading authors: Wolff, Blumenbach, and especially Herder from whom Kant borrowed the word. Herder had taken it from Wolff. This ques-tion concerns more generally the use of biological terminology to describe the epistemological field. In the "Methodology of Pure Reason" of the first *Critique*, which is the second part of the book, Kant describes the growth of rational knowledge with the terminology of the theory of germs,[75] hav-ing already spoken of *Keime* (germs) in the Transcendental Deduction. More generally, "organisms" are increasingly, even through the first *Critique*, a schema to understand what is meant by a rational realization.

In fact, a first reading indicates that the debate between preformationism and epigeneticism is used by Kant to exclude both a dogmatist (i.e., Leibniz-ian) conception of the categories (as contending that categories are from the beginning the general forms of the objects themselves) and Humean empiricism (as claiming that categories are abstracted from a form-free sense experience). In this case, however, one should wonder why Kant speaks of an *epigenesis* of pure reason, rather than trying to find an intermediate posi-tion between preformationism (which would correspond to dogmatism) and epigenesis (which would correspond to empiricism). Moreover, this commit-ment to an "epigenesis of pure reason" contrasts with what we saw of Kant's preference for a "generic preformationism" rather than crude epigenesis in the case of living beings: What justifies this difference in the two contexts?

To deal with this question, one has to make clear the meaning of Kant's terms *Keime* and *Anlage* in order to decide to what extent those concepts belong to a preformationist or to an epigeneticist framework. In "Preforming the Categories," Phillip Sloan investigates along these lines and argues that, contrary to previous interpretations, Kant's commitment to epigenesis allows him to think anew the issue of a priori knowledge while revising the first *Cri-tique* in 1787. This point can be conceived as correlated with the issue of *innate* knowledge within man as a living being. Here, attention to biological matters in Kant modifies in interesting ways the picture of the critical philosophy as an anti-naturalistic philosophy. The attempt to understand the obscure theory of an "original acquisition" of categories in the answer to Eberhardt could benefit from this new reading.[76]

1.6. The History of Kant's Thought

Of course, Kant's reflective positions on biology from 1755 to the end of his life evolved considerably, and not simply concerning the meaning and application of epigenetic conceptions. One historical topic is the fate of those issues in the emerging critical philosophy, given that in the so-called pre-critical writings, topics in the analysis of life sciences and references to authors in the fields (Wolff, Haller, Stahl, etc.) are numerous. One must not forget, in this context, that the first important writing on biological topics ("Von den veschiendene Menschenrassen," 1775) was written in a pure pre-critical fashion, since it belongs to a project of *Populärphilosophie*, and not at all to the program of a critique of metaphysics. This critical program is itself more metaphysical than the *Populärphilosophie*, which sounded rather like a pure rejection of metaphysics, like Reid's philosophy of common sense (or Reid as German thinkers such as Garve interpreted him).[77] Questions addressing Kant's evolution through those pre-critical writings to the critical works include of course the three Critiques and the question already mentioned concerning whether or not Kant made meaningful changes throughout the *Critiques* while thinking about biology.[78]

The other historical topic concerns the changes occurring in Kant's thought after the *Critiques*, and principally in the *Opus postumum*. Here, historians of philosophy disagree considerably concerning the real aim of this unfinished project, untitled by Kant, which is supposed to complete the passage from metaphysics to the philosophy of nature. Friedman and Mahieu offer some interpretations.[79] What is not controversial in contemporary literature is the fact that Kant's project, in the *Opus postumum*, stems from internal reasons, concerning the articulation of transcendental philosophy in the *Critique*, the metaphysics of nature, and the natural sciences[80]—as well as from external reasons, namely, the emergence of German idealism, and especially Fichte's transcendental philosophy, which appeared clearly too audacious to Kant. Despite his critics, Kant felt compelled to show that his own philosophy could answer to the challenges raised and apparently solved by Fichte.[81]

The first problem of the work is reminiscent of the problem of the *Critique of Judgment*, namely, how to get to the empirical and particular laws of nature, once we have a transcendental account of the general lawlikeness of nature (given in the first *Critique*). So it is a debate for the historians to understand the precise relationships between those two works, given that one of them is clearly unfinished.

These considerations imply that the whole meaning of Kant's theories in the life sciences after the *Critiques* is difficult to reconstruct—if such a reconstruction is even possible. Some striking contrasts have been noted: Kant seems more liberal with his own distinction between regulative and constitutive, here thinking of purpose as a constitutive principle;[82] in addition, he relates his concept of organism to an internal experience of living bodies,[83] which seems very far from the methods and spirit of the *Critiques*.[84] From this point of view, even the use of the word *Organismus* seems puzzling. This raises retrospectively a general problem in Kantianism: What is the exact relation among *Organisierte Wesen, Organismus*, and living being? Are we allowed to take for granted, in the *Critique of Judgment*, that those organized beings that are objects of the Kantian analyses are the same things as living beings? If this is not the case, how do we deal with this difference?[85]

This indicates that the whole status of those texts has to be appreciated and criticized if they are to be taken into account in a reconstruction of Kant's thinking about the life sciences. Nevertheless, they provide valuable materials concerning the difficulties and opacities inherent in Kant's analytics of biology and theory of germs, as well as concerning the general contexts—internal to Kant's thinking and in his debates with contemporaries—wherein those conceptions had to be used, as soon as a defense and self-understanding of the critical philosophy were undertaken by Kant.

This leads us directly to the second order of issues, namely, Kant's legacy and the definition and qualifying of a "Kantian tradition in biology" or a "vital-materialism" as termed by Lenoir.

2. Issues after and outside Kantianism

2.1. Regulative Principle: What Did That Mean and for Whom?

If the claim of natural purpose as a regulative principle is obscure in Kant, it was a fortiori difficult for Kant's readers to understand what is meant by this concept, even if they felt that they needed it in order to conceive what they were effectively doing in the rising field of biology.

A point of debate in the literature following Lenoir is the extent to which life scientists from Blumenbach to Von Baer respected the Kantian spirit. Lenoir contended that the alliance of teleology as regulative and the mechanism constituted the teleomechanist research program of the Göttingen school and, widely speaking, of German biology until Johannes Müller, even if the history of this school showed a progressive conflating of the two poles (constitutive/regulative).[86] He shows that this confusion, proper to life scientists and to *Naturphilosophie*, can be traced back to "generational factors" also active in political thinking—namely, the fact that, contrary to Kant and Blumenbach, Schelling, Keilmayer, and Link lived through the French Revolution and the decline of authoritarianism in their youth.[87] Larson has also claimed that what Kant took as regulative was more and more seen as constitutive by the biologists.[88] Richards in "Kant and Blumenbach: A Reactive Misunderstanding" and Caneva in "Teleology without Regrets" guess that from the beginnings this distinction was of no concern for the life scientists, even if they used the word. Jardine sees a shift between a Kant-Blumenbach set of questions for natural history and the later morphological and embryological problematics, but situates this shift in Kielmayer's discourse.[89]

Since the publication of Mendelsohn's and Temkin's papers and Rittersbuch's book, which gave some insightful reconstructions of German biology at those times, some historical elements for reconstituting German life sciences in the romantic era by the aforementioned authors as an incipient biology (with its constitutive use of what Kant called purposiveness) are to be found in works by Ilse Jahn, David Knight, and Nicholas Jardine.[90]

In fact, it should be noted, once again, that Kant's aim was at first a metaphysical one, and the distinction had its primary meaning in this context. But as Blumenbach and his followers had another aim, namely, explanation of living creatures, the distinction could appear less clear-cut to their eyes. What mostly impressed them in the Kantian distinction is a claim of irreducibility of life sciences using teleological judgment naturalistically. This is attested by Blumenbach's addition of the word *Zweckmässigkeit* in order to characterize his *Bildungstrieb*, when he revises the fifth edition of his *Handbuch der Naturgeschichte* following the publication of the third *Critique*. Kant's metaphysical point—that is, that this autonomy of the living realm is epistemological rather than ontological—is no longer salient in this perspective.[91]

2.2. Issues about Morphology in the Kantian Period

2.2.1. Morphology and Types. Taken as constitutive, the idea of natural purpose grounded the vindication of a new kind of science, dealing with a new kind of fact: the archetypes. Blumenbach interpreted the *Bildungstrieb*—the force driving epigenesis—as aiming to a specific goal characterizing the form of the species, which was the type. So the legacy of Kant's idea of original organization, epigenesis, germs, and so on and, mostly, reflexive judgment on natural purpose, allowed Blumenbach and his followers (Keilmayer, etc.) to formulate the methodology of a science of types and to justify its autonomy.

Of course, here this morphology is deeply entrenched with embryogenesis as a process aiming at the types. It is not surprising that later, in works by Geoffroy Saint-Hilaire and Meckel, morphology would become linked to embryology.[92] This idea of types surely preceded Kant,[93] but in forging the so-called Kantian tradition in biology, some support for this idea, and especially a legitimation for a methodology, was surely gained. Philip Rehbock analyzed this connection between the idea of natural purpose and morphology.[94] Morphology was developed by Goethe, in parallel with his study in botany; his discovery of an *Urpflanze* matches exactly the idea of *Urtypus* in comparative anatomy.[95] The balanced influence of Goethe and Kant is of course a major point of interest.[96] The school of transcendental morphology combined the heritage of Blumenbach and Kant with that of Goethe and his important follower Lorenz Oken, together with Geoffroy Saint-Hilaire's influence in the *Philosophie anatomique*.[97] Rehbock and Sloan show that this research tradition entailed major consequences in British natural science (Knox, Grant, Green, Owen).[98] For many reasons, Owen can be seen as the final point of this history, within which philosophy, embryology, compared anatomy, and morphology found themselves deeply entangled.

In *The Romantic Conception of Life*, Richards studies most of the authors involved in order to give a comprehensive reconstruction of what he calls the "romantic conception of life" as a self-structured whole, a real "conception" rather than a rhapsody of more or less fantastic theories. In this book two important issues following from the present paragraph are developed.

2.2.2. *Naturphilosophie* and the Kantian Research Program. It was a tradition in the history of science to undermine the significance of works from Schelling, Oken, and Hegel and generally of all works that are called Naturphilosophie.[99] *Naturphilosophie* has long been seen as an obscure metaphysics and an obstacle to the real science done by comparative anatomists from Cuvier to Owen or, in the second half of the nineteenth century, by physiologists, chemists, or naturalists such as Du-Bois Reymond and Schwann. This view is encouraged by the self-assessment of these authors, but it seems to neglect other historical facts, such as Owen expressing his great admiration for Oken.[100] The time has come for a reappraisal of *Naturphilosophie*, as is clearly illustrated by the aforementioned works of Rehbock, Sloan, or Richards.

Jardine defines the idea of *inquiry* and distinguishes the set of questions proper to a Kantian-Blumenbachian inquiry, and inquiry in the sense of Goethe, Oken, and their followers in *Naturphilosophie*.[101] The idea of type becomes salient and decisive for the latter. However, by conflating what Kant identified as regulative or as constitutive, Kielmayer, Reil, and other *Naturforscher* were approaching the metaphysical positions of this *Naturphilosophie*, since the latter were from the beginning built on a critique of Kant's dualism and a transformation of the regulative into something more constitutive than what appeared merely regulative to Kant.[102] For this reason, Lenoir's clear-cut demarcation between Kantian biology and *Naturphilosophie*

seems unlikely to withstand any further historical investigations. Nevertheless, recently Reill has argued that "a major difference separated Enlightenment vitalism [to which Blumenbach, Kant, and Kielmayer subscribe, and
which Reill has shown to be pervasive in the mid-eighteenth century life
sciences] and Naturphilosophie," namely, the tone of skepticism regarding any
universal explanation—a tone characteristic of the former and absent from
the scientific program of Schelling, Goethe, and others.[103] It may be that
this genealogical difference represents the most genuine contrast to be made
between Naturphilosophie and the previous developments in German biology
along Wolffian and Kantian lines.[104]

Of course, one general question emerging as soon as we take this Naturphilosophie seriously, and recognize that it was caught within the conceptual
foundations of what we call biology, is that our idea of science comes under
focus. If we inherited this idea from precisely those people who disregarded
Naturphilosophische ideals, and elaborated their own image as the opposite of
the discourse they caricatured, then re-evaluating real Naturphilosophie could
lead to a critique of our analytical, reductionistic, anti-holistic conception of
natural science. This direction of thought is clearly suggested in the works of
Jardine, Knight, and Richards.[105]

2.2.3. Transformism. The transition from German biology to British morphology proves to be of highest importance in the history of biology, as soon as we
recall that Owen was surely one of the main friends of and inspiration to Charles
Darwin. This shows that our topic is connected to the question of the emergence of evolutionism. Richards's latest books contend that, against the received
view, Darwinism has some filiations with German biology.[106] Although reluctant to subscribe to this view fully, Sloan emphasizes the filiations between German biology and the British comparative anatomy within which Darwinism
was elaborated.[107] Of course, this issue is related to the preceding one, since the
opposition of analytical natural science and obscure unscientific Naturphilosophie
has often been illustrated by the contrasts of Darwin and some of the German
transcendental morphologists. This relationship deeply challenges Lenoir's interpretation, which in the end tries to illuminate Von Baer's strange reluctance to
Darwinism through an appeal to a deep opposition of ontological and methodological frameworks between the two authors.[108]

The main point here refers to the idea of types and archetypes. After Goethe
and Blumenbach, German biologists shared the idea that living individuals,
during their embryogenesis, were driven by a building force toward a type that
is characteristic of their species. Cuvier's major "embranchements" were interpreted by Von Baer as various aims of this building force, and finally as kinds of
splitting directions of this force. This allowed him to reconcile Cuvier's view
with Geoffroy's opposite idea of a unity of type. Ospovat calls this conception a

"branching conception" and highlights its importance for Darwinism.[109] This indicates that the types, according to these authors, are related to one another by a kind of process.[110] (An important issue here is the case of the law of recapitulation, called "law of Meckel-Serres," of which a first formulation is to be found in Keilmayer's discourse.)[111]

Considering these elements, we might wonder whether those authors, along the lines sketched by Blumenbach and Kant, envisioned a transformation of types, even if those types were—as Von Baer puts it—some kinds of ideal form, close to Platonic essences.[112] It is not to be discussed that from Kielmayer on it was thought that the types could be ordered in a process and that this process was chronologically realized on earth.[113] The process itself, the transformation, however, was an ideal one. Even Herder thought that species, corresponding to the types, came to earth one after the other.[114] Nevertheless, the question is whether the transformation of types into other types is a kind of logical necessity, or is something which occurs empirically in the world. These issues seem to be unclear in the texts themselves, and for their authors.[115] In his *Romantic Conception*, Richards claims that a real transformism was conceived by some authors anyway and that this idea of transformation of types was of interest for Darwin.[116] Of course, many features of the Darwinian theory are very far from the German biologists and deeply anchored in the English intellectual world, with its long-standing habits of empiricism.[117] In any case, it seems certain that those "romantic" conceptions, however obscure they may appear to the contemporary life scientists themselves, had a strong impact on the gestation of evolutionary thinking.

2.3. Reductionism and the Life Sciences

If life scientists found in Kant those two important elements, namely, a claim of irreducibility for living organization and the use of a teleological framework, and if those two claims, however foundational they were for biology, appear to us as opposed to the scientific biology, which sees no discontinuity between macromolecules and living beings and admits no use of teleological principles and forces, the final issue is the relevance of those historical matters to the philosophical understanding of biology. Were all those people wrong? And if they were, why did those wrong beliefs lead to major achievements in the life sciences?

Lenoir's book tried to understand the rejection of *Naturphilosophie* and teleomechanism by the generation of Du Bois Reymond, Schwann, and others. He found some historical relationships between the former research program and the latter scientists. Settling this issue decisively appears to be of interest if one wants to explain the strength and the relevance of teleological thinking for the concrete achievements of the German biological school.

Historically, of course, the preceding issue of transformism is included in this general topic. Moreover, dealing with this question would lead us to define two contrasting images of scientific work, to state their historical relationships, and then to determine their real bearing on concrete scientific activity, as ideals vindicated by the scientists themselves. In fact, such an issue proves the relevance of our historical reconstruction for the philosophy of science. This relevance is recently attested by some works in the literature on functions in biology, such as Matthew Ratcliffe's "The function of functions," which advocates a Kantian position in order to provide a unified framework to account for functional explanations in biology, or Peter McLaughlin's *What Function Explains* (2001), which contends that any theory of functions has to be committed to an ontology where there are, besides the categorical properties of physical objects and propositional attitudes, some self-reproducing entities, to which alone functional ascriptions could be applied—a thesis clearly evoking a strictly Kantian inspiration. At first blush, reductionism and the explanatory specificity of biology are at stake here.

3. The Contributions of This Volume

This short survey illustrates to what extent the issues related to "Kant and biology" covered by our volume could be of interest for general questions in history of philosophy, as well as history of biology and philosophy of science. I will now indicate the precise topics treated in the chapters and point out the articulation of the book in order to summarize some of the novel perspectives brought forth here. It should be noted that such a collection could also be of interest due to the sometimes conflicting points of view elaborated by the authors.

3.1. Kant, Embryology, and Epigenesis

To have a clear idea of Kant's thinking on biology, it is of the utmost importance to understand what "epigenesis" means for him, and then, in order to decide the originality of Kant's theory and relate it to his contemporaries, Haller, Wolff, Blumenbach, it is also important to undertake a survey of the various versions of epigenesis in the eighteenth century. A few chapters in this book bear on this question. Jean-Claude Dupont provides an important and original illumination here, by tracing it back precisely to Wolff's original and strongest formulation of a theory of epigenesis and by stating its challenges and novelties. A focus on Wolff's treatise on the formation on the intestine is necessary to understand what was needed for epigeneticism to overcome preformationism, delineating thereby the framework of the methodological and epistemological discussions on life sciences at the time.

Epigenesis is also at the center of Huneman's and Fisher's chapters, as well as Zammito's. These three texts take a more internal perspective on Kant. Zammito argues that Kant was not committed to what was salient for his contemporaries in epigenesis, namely, hylozoism, and used the concept to draw limits around it. This makes clear a connection between the metaphysical concerns about limits of mechanism in nature—and hence, limits of the transformations of species undergone through such mechanism—and the attention to embryology.

Huneman analyzes the place of Wolffian epigenesis in the internal shift between the first and the third *Critique*, concerning the meaning of purposiveness, which has successively been described as a principle of reason and then as the principle of the power of simply reflexive judgment. He argues that this shift in Kant's thought can partly be illuminated by reconstituting the constraints of an account of his own commitments concerning the theory of generation, and hence of the Wolffian embryology as the main matrix of such commitments.

3.2. Natural History, Classification, Comparative Anatomy

Mark Fisher pays careful attention to Kant's commitment to the theory of epigenesis in the context of a question concerning the meaning of "natural history." He situates this project in the contexts of anthropology and physical geography in Kant's thinking, using unpublished materials from the *Akademie Ausgabe*. He finally establishes a strong connection between Kant's version of epigenesis (the so-called theory of "generic preformationism") and the novel framework he established, in terms of a history of nature, for the task of classification.

The question of natural history and taxonomy is in the focus of the final three chapters, which address Kant's legacy for subsequent life scientists. Stéphane Schmitt argues that the idea of Kantian biology has to be revised in the light of the works on classification by the so-called teleomechanists, because this classification also uses French natural history, and the quite complex theoretical use of the concept of biological forces elaborated there. Schmitt points out that Keilmayer's general idea of drawing an entire classification of animals through the presence or lack of the biological forces in the sense of Blumenbach, and then the subsequent shift from this attempt to a classification using organs instead of forces, already found resources in some previous concepts, among others, those of Vicq d'Azyr's anatomy. Accordingly, there appeared progressively a novel continuity across French anatomy, teleological thinking, and *Naturphilosophie*—namely, Vicq d'Azyr, Blumenbach, Oken—by which the historian can challenge the previously accepted borders of those discourses.

While Robert Richards reconstructs the connections between teleology and Goethe's morphological thinking, which is of primary importance for both *Naturphilosophers* and transcendental morphologists, his chapter contests another well-established frontier; namely, that between poetry and science. This challenge is deeply entrenched in the project of science envisioned by Goethe and, later, Schelling. There appears to be a connection between Kant and *Naturphilosophie*, through the third *Critique* considered in the unity of its two parts, which is what attracted Goethe the most in this work. Hence, critical notions of purposiveness that could be conceived as necessary with regard to the mere context of a science of nature played an important role, once extracted from this context, in the anti-Kantian advocacy of the possibility of intellectual intuition.

In his chapter, Phillip Sloan stresses the importance of these connections and research projects on British comparative anatomy. Here, Green's work shows that, once again, teleological thinking in a Kantian manner and a *Naturphilosophische* spirit in morphology were unified in the anatomical tradition of John Hunter in order to provide a new framework whose principal heir was Green's successor and Darwin's mentor, Owen. This process relied heavily on the key figures of Schelling in Germany, and then Coleridge's interpretation of Schelling's philosophy of nature in the English context. Sloan encourages us to pay attention to the productive misreading of authors by the subsequent generation. Here, it is a Schellingian Kant—which is, historically, a massive mistake—who allowed the British morphologists to formulate their project.

The contrasts and differences of these final three chapters, nevertheless, stress the importance of hybrids stemming from distinct research programs and commitments, in the first period of the nineteenth-century life sciences. Such crossings proved highly fruitful in the definition of a framework for the life sciences within which a modern schema of biological disciplines emerged— namely, physiology, embryology, morphology, and comparative anatomy, both of which involve significant cross-references to embryology. The authors also emphasize the fact that elements of some varied teleomechanistic strategies were involved in those mixtures in order to be corrected, rectified, or falsified. Thus, they show the significance of these strategies in this peculiar history.

Notes

1. *KU*, §65, Ak. 5, 372.
2. *KU*, §65, Ak. 5, 373. For two recent accounts of the thesis of "natural purpose" in Kant's theory of organism and its articulation to mechanical or nomological explanation, see Hannah Ginsborg, "Kant on Understanding Organisms as Natural Purposes," in *Kant and the Sciences*, ed. Eric Watkins, 231–59, and Philippe Huneman, *Métaphysique et biologie. Kant et la constitution du concept d'organisme* (Paris: Puf, forthcoming).

3.   See the famous sentence in §81 of the *Critique of Judgment*: "Regarding the theory of epigenesis, no one made more contributions than the councillor Blumenbach, both concerning the evidences in its favour, and the foundations of the principles of its application" (Ak. 5, 424).

4.   Occurrences in Ak. Ag. XIV–XV, and *Anthropologie in der pragmatischer Hinsicht*.

5.   Kant wrote to Soemmering an important letter about his treatise on localizing the organ of the soul. See Luigi Marino, "Soemmering, Kant and the Organ of the Soul," in *Romanticism in Science: Science in Europe, 1790–1840*, ed. Stefano Poggi and Maurizio Bossi (Dordrecht: Kluwer, 1994), 47–74, which contains a nice survey of the state of the fields in physiology. See also Peter McLaughlin, "Soemmering und Kant: Uber das Organ der Seele und den Streit der Fakultäten," *Soemmering-Forschungen* 1 (1985): 188–95.

6.   *Anfangsgründe der Naturlehre* (Göttingen, 1784).

7.   The *Traüme* is a good example of such an abundance of references to those four authors. On Kant's scientific background, much information can be found in Erich Adickes, *Kant als Naturforscher* (Berlin: De Gruyter, 1923), vol. 2, and Hans-Joachim Lieber, "Kants Philosophie des Organischen und die Biologie seiner Zeit," *Philosophia naturalis* 1 (1950): 553–70. Some indications are to be found in John MacFarland, *Kant's Concept of Teleology* (Edinburg: Edinburg University Press, 1970), and in Reinhard Löw, *Philosophie des Lebendigen. Der Begriff der Organischen bei Kant, sein Grund und seine Aktualität* (Frankfort: Suhrkamp, 1980).

8.   *KU*, §80, Ak. 5, 417ff.

9.   *"Ursprüngliche Organisation"*: *KU*, §80, Ak. 5, 418, §81, Ak. 5, 424, etc.

10.   About this, see Marc Klein, *Regards d'un biologiste* (Paris: Hermann, 1980), Giulio Barsanti, "La naissance de la biologie. Observations, théories, métaphysiques en France, 1740–1810," in *Nature, Histoire, Société, Mélanges offerts à Jacques Roger*, ed. Roselyne Rey et al. (Paris: Klincksieck, 1995), 196–228, and "Lamarck and the Birth of Biology. 1740–1810," in *Romanticism in Science: Science in Europe, 1790–1840*, ed. S. Poggi and M. Bossi (Dordrecht: Kluwer, 1994), 47–74, and Philip C. Rittersbuch, *Overtures to Biology: The Speculations of Eighteenth-Century Naturalists* (New Haven: Yale University Press, 1964).

11.   On the pervasive influence of historical thinking in nineteenth century, see William Coleman, *Biology in the 19th Century: Problems of Form, Function and Transformation* (Cambridge: Cambridge University Press, 1979).

12.   *Theoria generationis* (Halle, 1759).

13.   *Entwicklungsgechichte der Thiere* (Göttingen, 1828).

14.   A benchmark here has been the publication of Plattner, *Menschenkunde für Ärtze und Weltweise* (1772).

15.   On those authors and their filiations, see Bernard Balan, *L'ordre et le temps* (Paris: Vrin, 1979); on their acquaintance to transformism, see Coleman, *Biology in the 19th Century*.

16.   *KU*, Introduction, Ak. 5, 180ff.

17.   "Hence (medical) physiology extends its very limited empirical acquaintance with the ends served by the structure of an organic body through a principle prompted merely by pure reason—and this to such an extent that it is assumed confidently, and with the agreement of all who understand the matter, that everything

in an animal has its utility and good aim" (*Critique of Pure Reason*, hereafter *KRV*, A688/B716).

18. On Kant's dialectics and systematicity, see Philip Kitcher, "Projecting the Order of Nature," in *Kant's Philosophy of Physical Science*, ed. Robert E. Butts (Dordrecht: Reidel, 1986), 201–35; Jonathan Bennett, *Kant's Dialectic* (Cambridge: Cambridge University Press, 1974); William Walsh, *Kant's Criticism of Metaphysics* (Edinburg: Edinburg University Press, 1975); Peter F. Strawson, *The Bounds of Sense: An Essay on Kant's* Critique of Pure Reason (London: Routledge, 1995), 226–31; Gerd Buchdahl, "Causality, Causal Laws and Scientific Theory in the Philosophy of Kant," *British Journal for Philosophy of Science* 16 (1965): 186–208, and *Metaphysics and the Philosophy of Science* (Cambridge: Belknap Press, 1969); Michael Friedman, "Causal Laws and the Foundations of Natural Science," in *Cambridge Companion to Kant*, ed. Paul Guyer, 186–90; Paul Guyer, "Reason and Reflective Judgment: Kany on the Significance of Systematicity," *Noûs* 24, no. 1 (1990): 17–43; and Henry Allison, *Kant's Transcendantal Idealism* (Yale: Yale University Press, 2003).

19. *KRV* (A653/B681), "First introduction" to the *Critique of Judgment*, Ak. XX, 232. Commentary in Gérard Lebrun, *Kant et la fin de la métaphysique* (Paris: Armand Colin, 1970), ch. 10.

20. "Considering the empirical truth, we must presuppose the systematic unity of nature as objectively valid and necessary" (*KRV*, A654/B682).

21. *KU*, Introduction, 181: "So ist [. . .] das Prinzip der Zweckmässigkeit der Natur (in der Mannigfaltigkeit ihrer empirischen Gesetze) ein transzendantales Prinzip."

22. Kitcher, "Projecting the Order of Nature."

23. First introduction of the *Critique of Judgment*, Ak. XX, 215.

24. This attests that Kant's solution to the problem of induction lies here rather than in the *Analogie* of experience.

25. *Teutsche Merkur*, 57–86, 1786, pp. 150–66. On Forster, see Robert Bernasconi (ed.), *Concepts of Race in the Eighteenth Century* (London: Thoemmes Continuum, 2001); Frederick Beiser, *The Fate of Reason* (Chicago: University of Chicago Press, 1990); and Ilse Jahn, "Georg Forsters Lehrkonzeption für eine 'Allgemeine Naturgeschichte' (1786–1793) und seine Auseinandersetzung mit Caspar Friedrich Wolffs 'Epigenesis'-Theorie," *Biologisches Zentralblatt* 114 (1995): 200–206.

26. On this theory, see, among many papers, Phillip Sloan, "Preforming the Categories: Eighteenth-Century Generation Theory and the Biological Roots of Kant's A Priori," *Journal of the History of Philosophy* 40, no. 2 (2002): 229–53; Zöller, "Kant on the Generation of Metaphysical Knowledge," in *Kant: Analysen—Probleme—Kritik*, ed. Hariolf Oberer and Gerhardt Seel (Wurtzburg: 1988), 71–90; Clark Zumbach, *The Transcendant Science: Kant's Conception of Biological Methodology* (Den Haagen: Martinus Nijhoff, 1984), 109; and Timothy Lenoir, *The Strategy of Life: Teleology and Mechanism in Nineteenth Century German Biology* (Dordrecht: Reidel, 1982), ch. 1.

27. Hannah Ginsborg draws some important consequences from this difference, since the teleological character of natural products is not a non-machine-like character; she is led to challenge McLaughlin's understanding of mechanism as a parts-determining-the-whole relationship ("Two Kinds of Mechanical Inexplicability in Kant and Aristotle," *Journal of the History of Philosophy* 42, no. 1 (2004): 33–65).

28. *KU*, §65.

29. On this theory and its relationship to adaptation, see particularly chapter 4 in this volume; Philippe Huneman, "Espèce et adaptation chez Kant et Buffon," in *Kant et la France-Kant und Frankreich*, ed. Robert Theis et al. (Hildesheim: Olms, 2005), 107–20.

30. Tournefort, *Eléments de botanique* (Paris, 1694); Linnaeus, *Systema naturae* (1735); Adanson, *Famille des plantes* (Paris, 1763); Antoine-Laurent de Jussieu, "Exposition d'un nouvel ordre de plantes" (1763); *Genera plantarum* (Paris, 1789). On natural history in this period, see Phillip R. Sloan, "Natural History, 1670–1802," in *Companion to the History of Modern Science*, ed. R. C. Olby, G. N. Cantor, J. R. R. Christie, and M. J. S. Hodge (London: Routledge, 1990), 295–313; and Nicholas Jardine et al. (eds.), *Cultures of Natural History* (Cambridge: Cambridge University Press, 1996).

31. On ontology and natural history, see Sloan, "Natural History, 1670–1802," and Phillip Sloan, "The Gaze of Natural History," in *Inventing Human Sciences*, ed. Christopher Fox et al. (Berkeley: University of California Press, 1995), 112–53.

32. "Discours sur la manière d'étudier l'histoire naturelle," *Histoire naturelle* (Paris: Imprimerie Royale, 1749), vol. 1, pp. 3–64; Jacques Roger, *Buffon: un philosophe au jardin du Roi* (Paris: Fayard, 1989); Phillip R. Sloan, "Buffon, German Biology, and the Historical Interpretation of Biological Species," *British Journal for the History of Science* 12 (1979): 109–53.

33. On Buffon's concept of species, see Paul Farber, "Buffon and the Concept of Species," *Journal of the History of Biology* 5 (1972): 259–84: Phillip Sloan and John Lyon, *From Natural History to the History of Nature* (Notre Dame: University of Notre Dame Press, 1981): Giulio Barsanti, "Buffon et l'image de la nature: de l'échelle des êtres à la carte géographique et à l'arbre généalogique," in *Buffon 88*, ed. Jean Gayon (Paris: Vrin, 1992), 255–95; and, criticizing Barsanti, Thierry Hoquet, "La comparaison des espèces. Ordre et méthode dans l'*Histoire naturelle* de Buffon," *Corpus. Revue de philosophie* 43 (2003): 355–416. On Buffon and Kant, see Jean Ferrari, "Kant lecteur de Buffon," in *Buffon 88*, ed. Jean Gayon, 155–62. In "Buffon, German Biology, and the Historical Interpretation of Biological Species," 116, Sloan points out a Leibnizian inspiration in Buffon's insistence on history rather than morphology and similarity in the definition of species and the method of taxonomy (inspiration supported by the French approval of Leibniz in the intellectual circles of Voltaire and the marquise du Châtelet). This source illuminates the fact that Kant, as a former disciple of Christian Wolff, was so receptive to Buffon's epistemological thesis on natural history. The German reception of Buffon then shifted around 1770: previously, he was seen as a misguided critic of the correct Linnean methodology in natural history.

34. On Girtanner, see Timothy Lenoir, "Generational Factors in the Origin of *Romantische Naturphilosophie*," *Journal of the History of Biology* 11, no. 1 (1978): 74–80, and Sloan, "Buffon, German Biology, and the Historical Interpretation of Biological Species," 137–42. Sloan contends that Girtanner was the first to promote a synthesis between Kant's idea of natural history as historical science and Blumenbach's notion of *Bildungstrieb*. Concerning the concept of species, this inspiration had been pushed to term by Illiger, *Versuch einer systematischen vollständigen Terminolgie für das Thierreich und Pflanzenreich* (Helmstadt, 1800).

35.   On geography, see Theodore Mischel, "Kant and the Possibility of a Science of Psychology," in *Kant Studies Today*, ed. Lewis W. Beck (Lasalle, IL: Open Court, 1969); chapter 4 in this volume; and James May, *Kant's Concept of Geography and Its Relation to Recent Geographical Thought* (Toronto: University of Toronto Press, 1970). Kant had been teaching geography since 1757.

36.   On the status of this distinction and its fate by and beyond Kant's thinking, see chapters 4 and 7 in this volume.

37.   For an old but interesting account of this theme in Kant's thinking, see Paul Menzer, *Kants Lehre von der Entwicklung in Natur und Geschichte* (Berlin: Reimer, 1911).

38.   Pierre Jean Georges Cabanis, *Rapports du physique et du moral de l'homme* (Paris, 1801); Jules-Joseph Virey, *Histoire naturelle du genre humain* (Paris, 1817).

39.   On the major European themes of the eighteenth-century anthropology seen from Scotland, see Paul B. Wood, "The Science of Man," in *Cultures of Natural History*, ed. Nicholas Jardine et al. (Cambridge: Cambridge University Press, 1996), 197–210. In general, see especially *Inventing Human Sciences*, ed. Christopher Fox et al. (Berkeley: University of California Press, 1995), and Wolff Lepenies, *Das Ende der Naturgeschichte: Wandel kulturellen Selbstverständlichkeiten in des Wissenschaften des 18. Und 19. Jahrhundert* (Munich: Hansen, 1976).

40.   Mareta Linden, *Untersuchungen zum Anthropologiebegriff der 18. Jahrhundert* (Bern: Herbert Lang, 1979). Robert Wokler argues that the question of the difference between man and animal gave rise to the criteria of the differences between races. "With the orangutan now cast out of the human race on the account of physical or material divergences only, the anthropologists came to be refocused on the apparent boundaries within the human race, rather than upon the animal limits of humanity" ("Anthropology and Conjectural History in the Enlightenment," in *Inventing Human Sciences*, ed. Christopher Fox et al., 45). This field of questions has been substituted for the previous natural history of man as "conjectural history" of the progress from natural and physical man to civilized and moral man, as practiced by Rousseau, Lord Monboddo, and Buffon, and requiring some "speculative and even fabulous classifications of the varieties of man."

41.   On constituting an anthropological discourse in Germany, especially by physicians (e.g., Plattner), see Linden, *Untersuchungen zum Anthropologiebegriff der 18. Jahrhundert*; Roy Porter, "Medical Science and Human Science," in *Inventing Human Sciences*, 53–87; and John Zammito, *Kant, Herder and the Birth of Anthropology* (Chicago: Chicago University Press, 2002), ch. 6 and 251–53 on Plattner.

42.   On Blumenbach's anthropology, see Stefano Fabbri Bertoletti, "The Anthropological Theory of Johann Friedrich Blumenbach," in *Romanticism in Science: Science in Europe, 1790–1840*, ed. Stefano Poggi and Maurizio Bossi (Dordrecht: Kluwer, 1994), 103–25. Blumenbach in the successive editions of his *Handbuch der Naturgeschichte* did not wholly accept Buffon's and Kant's definition of races and species in historical terms (Sloan, "Buffon, German Biology, and the Historical Interpretation of Biological Species," 136–37); the application of those terms had to wait for Girtanner (1796).

43.   Krüger, *Naturlehre* (1740–49) and *Versuch einer experimental Seelenlehre* (1756); Unzer, *Philosophische Betrachtungen des menschlichen Körper überhaupt* (1750).

See also Gary Hatfield, "Remaking the Science of Mind: Psychology as Natural Science," in *Inventing Human Sciences*, 184–232, for the general strategy of those authors. Zammito writes perspicuously: "Anthropology as a new discourse in Germany was the product of these 'philosophical physicians' with the popular 'philosophers for the world'" (*Kant, Herder and the Birth of Anthropology*, 245).

44.   On this crucial question concerning Kant's system, see Mischel, "Kant and the Possibility of a Science of Psychology"; Gary Hatfield, "Empirical, Rational and Transcendental Philosophy: Psychology as Science and as Philosophy," in *Cambridge Companion to Kant*, ed. Paul Guyer, 200–228; Thomas Sturm, "Kant on Empirical Psychology: How Not to Investigate the Human Mind," in *Kant and the Sciences*, ed. Eric Watkins, 163–84; and Rudolf Makkreel, "Kant on the Scientific Status of Psychology, Anthropology and History," in ibid., 185–201.

45.   "Originally, there was a unique species of men, which, having multiplied all over the Earth, underwent various changes under the influence of the climate, of the food, of different ways of life, of epidemias and also through the infinitely varied mixing of more and less similar individuals" ("Il n'y a eu originairement qu'une seule espèce d'hommes, qui s'étant multipliée et répandue sur toute la surface de la terre, a subi différens changemens par l'influence du climat, par la différence de la nourriture, par celle de la manière de vivre, par les maladies épidémiques, et aussi par le mélange varié à l'infini des individus plus ou moins ressemblans", "Variétés dans l'espèce humaine," *Histoire naturelle*, vol. 3, pp. 371–530). On Buffon and human species, see Roger, *Buffon*; Sloan and Lyon, *From Natural History to the History of Nature*; and Michèle Duchet, *Anthropologie et histoire au siècle des Lumières* (Paris: Albin Michel, 1995). On polygenism and monogenism, see Wokler, "Anthropology and Conjectural History in the Enlightenment," 35–38; and Sloan, "Buffon, German Biology, and the Historical Interpretation of Biological Species," 124–25; whereas the Buffonian definition of species yielded monogenism, Buffon's own doubts on the applications of his principle in the subsequent volumes of his *Histoire naturelle* authorized a reaffirmation of polygenism, especially reflected in the influential book by Lord Kames (1774), which was translated into German the following year.

46.   On this episode, see, e.g., Sloan, "Buffon, German Biology, and the Historical Interpretation of Biological Species," 131–33; and Susan Meld Shell, *The Embodiment of Reason: Kant on Spirit, Generation and Community* (Chicago: University of Chicago Press, 1996), 196–201.

47.   For a survey of the problems of races in the eighteenth century, see Bernasconi, *The Concept of Race*; and Robert Bernasconi and T. L. Lott (eds.), *The Idea of Race* (Indianapolis: Hackett, 2000), where important materials are presented.

48.   However, this does not affect the fact that purposiveness requires some principles other than those of lawlikeness to be grounded. On natural purpose in eighteenth century, its critiques by Hume in the *Dialogues on Natural Religion* and by Kant, and the context of natural theology, see Phillip Sloan, "The Question of Natural Purpose," in *Evolution and Creation*, ed. Ernan MacMullin (Notre Dame: University of Notre Dame Press, 1983), 121–50.

49.   Lawlikeness, according to Buchdahl, is founded not by the understanding but by reason with its principles: "The second analogy does not show that nature is lawlike, but that the concept of law is built into our notion of each objective element of nature"

("Causality, Causal Laws and Scientific Theory in the Philosophy of Kant," 201). Kenneth Westphal in *Kant's Transcendental Proof of Realism* (Cambridge: Cambridge University Press, 2004), and Kitcher in "Projecting the Order of Nature," agree on this insufficiency of the Analogies regarding the establishment of a meaningful experience and a satisfactory rational knowledge of nature. Guyer ("Reason and Reflective Judgment: Kany on the Significance of Systematicity," *Noûs* 24, no. 1 (1990): 17–43) and Allison, *Kant's Transcendantal Idealism*, are less emphatic but agree on a weaker form of insufficiency. However, Michelle Grier gives the most extended account of the regulative function of reason, making sense of the coexistence in Kant of both seemingly incompatible assumptions, that reason carries a transcendental illusion which is inevitable and that the metaphysical fallacies relying on this illusion have to be criticized. The illusion itself, as inevitable, finds its proper and positive meaning as a regulative principle (*Kant's Doctrine of Transcendental Illusion* (Cambridge: Cambridge University Press, 2001), 263–301).

50.   For example: "an analogy of experience is only a rule according to which the unity of experience (and not the perception itself, as empirical intuition in general) must result from perceptions, and applies to objects (to phenomena), not as a *constitutive* principle, but only as a *regulative* principle" (*KRV*, Ak. 3, 161). Commentary in Norman Kemp-Smith, *A Commentary to Kant's* Critique of Pure Reason (London: Macmillan, 1923), 356ff.; see the interpretation of Jules Vuillemin, *Physique et métaphysique kantiennes* (Paris: Puf, 1987), ch. 3. Kant himself indicates the equivocal meaning of the constitutive/regulative distinction (*KRV*, A664/B692).

51.   On Kant's conception of methodology concerning organisms, compared with his general conception of systematicity, see Zumbach, *The Transcendant Science*, and Renate Wahsner, "Mechanism—Technizism—Technizism—Organism: Der epistemologische Status der Physik als Gegenstand von Kants *Kritik der Urteilskraft*," in *Naturphilosophie in Deutschen Idealismus*, ed. Karen Gloy and Paul Burger (Stuttgart: Fromann-Holzboog, 1993), 1–23.

52.   Such a reading is, for instance, to be met in Lebrun, *Kant et la fin de la métaphysique*, or Kitcher, "Projecting the Order of Nature."

53.   Of course, this issue on the meaning of "regulative" is related to an interpretative decision concerning the differences between the two main texts dealing with the order of nature, as we saw it: introduction of the *Critique of Judgment* and the Appendix to the transcendental dialectics. McLaughlin (*Kant's Critique of Teleology in Biological Explanation: Antinomy and Teleology* (Lewiston: E. Mellen Press, 1990)), for example, considers that there is no major change between the two texts. Buchdahl ("Causality, Causal Laws and Scientific Theory in the Philosophy of Kant," 205–7) examines on the same level the Appendix and the *Critique of Judgment*; for him, the "reflexive judgment" of the latter equates the "hypothetical use of reason" in the former (p. 202). See also Béatrice Longuenesse on the continuity of Kant's conception of judgment and on the importance of characterizing the teleological judgment as "*simply* reflexive judgment" (*Kant et le pouvoir de juger* (Paris: Puf, 1993), 209–11).

54.   In "Causality, Causal Laws and Scientific Theory in the Philosophy of Kant," Buchdahl addresses the principle of causality in the 2nd Analogy. The rule, according to which any change has to be conceived—as it is written in the Analogy (A189–B232)—does not mean a general principle of causality in the world;

therefore, he concludes that, since it has only to constitute objectivity of events, it cannot ensure the lawlikeness of events, which has to be grounded on a regulative principle of reason. In "Causal Laws and the Foundations of Natural Science," 170–86, Michael Friedman contests such interpretation, suggesting that "all empirical judgments are ultimately to be grounded in transcendental principles for Kant" (p. 186). Finally, in *Kant's Transcendental Proof of Realism*, Westphal establishes that the Analogies as a whole cannot rule out the case of internal causes bringing about changes, and hence are not the complete sufficient principles of any natural science.

55.   McLaughlin, *Kant's Critique of Teleology in Biological Explanation*, 141ff.

56.   Heinrich Cassirer, *A Commentary of Kant's* Critique of Judgment (London: Methuen, 1938), 343ff. Kant "should have said" that mechanism here means determinative judgment, whereas we are in the context of a critique of reflexive judgment (p. 351).

57.   *KRV*, A687/B715.

58.   For a general account of the nature and role of the antinomies in the Kantian system, see Strawson, *Bounds of Sense*, 193ff.; Victoria Wike, *Kant's Antinomies of Reason: Their Origin and Their Resolution* (Washington: University Press of America, 1982); and Grier, *Kant's Doctrine of Transcendental Illusion*. Huneman, in chapter 3 in this volume, develops the difference between the Antinomy in the third *Critique* and the relation of mechanism and purposiveness as they are defined in the first *Critique*.

59.   For a classical discussion of this antinomy, see Strawson, *Bounds of Sense*, 206–19. Alfred Ewing provides an interpretation of the reconciliation (*Kant's Treatment of Causality* (Oxford: Archon Books, 1924), 231–35). Alison, in *Kant's Transcendental Idealism*, argues that any conciliation has to rely on transcendental idealism.

60.   Wahsner, "Mechanism—Technizism—Organism," assimilates the two antinomies. McLaughlin, in contrast, notices that whereas the first *Critique* uses the difference before/after in order to conceive mechanism, the third *Critique* adds the category of parts/whole, which is distinctive of its antinomy of teleology (*Kant's Critique of Teleology*, 154). MacFarland notices that the third *Critique's* antinomy is solved by a common reference to the supersensible, whereas the third antinomy in the *Critique of Pure Reason* is solved by opposing the freedom's reference to supersensible, with the natural causality's attachment to the sensible realm (*Kant's Concept of Teleology*, 170). For this reference to supersensible in those two antinomies, see also McLaughlin, *Kant's Critique of Teleology*, 178.

61.   George Schrader's important paper on teleology is illuminating on this issue ("Status of Teleological Judgment in the Critical Philosophy," *Kant-Studien* 45 (1953–54): 203ff.). See also Cassirer, *A Commentary*; Robert Butts, "Kant's Schemata as Semantics Rules," in *Kant Sstudies Today*, ed. Lewis W. Beck (Lasalle, IL: Open Court, 1969), 290–300; and Reinhardt Brandt, "Analytic/ Dialectic," in *Reading Kant*, ed. Eva Schaper and Wilhelm Vossenkuhl (Oxforrd: Oxford University Press, 1989).

62.   Lebrun, *Kant et la fin de la métaphysique*; MacFarland, *Kant's Concept of Teleology*; Schrader, "Status of Teleological Judgment in the Critical Philosophy"; and Giorgio Tonnelli, "Von den verschiedenen Bedeutungen des Wortes Zweckmässigkeit in der Kritik der Urteilskraft," *Kant-Studien* 49 (1957/58): 154–66. One of the

oldest investigations is Konrad Marc-Wogau, *Vier Studien zu Kants Kritik der Urteilskraft* (Uppsala: Uppsala Universitets Årsskrift, 1938).

63. *KU*, §65, Ak. 5, 372.

64. "Von den verschiedenen Menschenrassen"; *KU*, §81.

65. On this debate, see Shirley Roe, *Matter, Life and Generation: Eighteenth Century Embryology and the Haller-Wolff Debate* (Cambridge: Cambridge University Press, 1980) and "Rationalism and Embryology: Caspar Friedrich Wolff's Theory of Epigenesis," *Journal of the History of Biology* 12, no. 1 (1979): 1–43; Jacques Roger, *Les sciences de la vie dans la pensée française au 18ème siècle* (Paris: Albin Michel, 1993); and Michael H. Hoffheimer, "Maupertuis and the Eighteenth Century Critique of Pre-existence," *Journal of the History of Biology* 15, no. 1 (1982): 119–44. While Peter Bowler ("Preformation and Pre-existence in the Seventeenth Century: A Brief Analysis," *Journal of the History of Biology* 4, no. 2 (1971): 221–44) illuminates the previous rise of preformationism, Charles Bodemer ("Regeneration and the Decline of Preformationnism in Eighteenth Century Embryology," *Bulletin of the History of Medicine* 38 (1962): 20–31) gives a meaningful account of the rise of epigenesis related to the impact of Trembley's discovery of the regenerating polyp (1744; on this event, see Marc Ratcliff, "Abraham Trembley's Strategy of Generosity and the Scope of Celebrity in the Mid-Eighteenth Century," *Isis* (2006)). Carlos Lopez-Beltram states a survey of the ideas on heredity at this period, which is useful to understand the various positions on generation ("Natural Things and Non-natural Things. The Boundaries of the Heredity in the 18th Century," in *A Cultural History of Heredity: 17th and 18th Centuries* (Berlin: Max Planck Institut, Preprint 222, 2002)).

66. For example, Lenoir, *Strategy of Life*, and "The Göttingen School and the Development of Transcendental *Naturphilosophie* in the Romantic Era," *Studies in the History of Biology* 5 (1981): 111–205; Kenneth Caneva, "Teleology with Regrets," *Annals of Science* 47 (1990): 291–300; Nicholas Jardine, *Scenes of Inquiry: On the Reality of Questions in the Sciences* (Oxford: Clarendon Press, 1991); Robert J. Richards, "Kant and Blumenbach on the *Bildungstrieb*: A Historical Misunderstanding," *Studies in History and Philosophy of Biological and Biomedical Sciences* 31, no. 1 (2000): 11–32; and John Zammito, *The Genesis of Kant's* Critique of Judgment (Chicago: Chicago University Press, 1992).

67. *KU*, §72, Ak. 5, 391. See also *First Principles of a Metaphysics of Nature*, Ak. 4, 544: "On the law of inertia (besides the one of the permanence of the substance) lies entirely the possibility of a science of nature. The opposite of it, and hence the death of any philosophy of nature, would be *hylozoism*." On the necessity of precluding any hylozoism for founding science, see Westphal, *Kant's Transcendantal Proof*, 164–66 and 201–8.

68. Peter Hans Reill, "Analogy, Comparison and Active Living Forces: Late Enlightenment Responses to the Critiques of Causal Analysis," in *The Sceptical Tradition around 1800*, ed. Karl van der Zande and Richard Popkins (Dordrecht: Kluwer, 1998), 207. In "Anti-mechanism, Vitalism and Their Political Implications in Late Enlightened Scientific Thought," *Francia*, Band 16/2 (Sigmarigen: Jan Thorbeck Verlag, 1990), 195–212, he sketched the political context and major implications of this intellectual shift, which entailed a critique of monarchy and sustained freedom of individual choice.

69. Alfred Baümler, *Das Irrationätsproblem in der Ästhetik und Logik des 18. Jahrhunderts bis zur Kritik der Urteilskraft* (Halle: Niemeyer, 1923), made a description of this kind of vitalism, within which emerges the problem of the power of rationality, in the background of the third *Critique*. Leskes (*Anfängsgründe der Naturgeschichte*, 2nd ed., 1784) posits an internal force within organisms; Beckmann (*Anfängsgründe der Naturhistorie*, 1764) ascribes to the constitution of organisms an ability to grow. Adickes (*Kants als Naturforshcer*, 479ff.) documented this literature on vital forces, which was quite contemporary to the third *Critique*. In *The Genesis*, Zammito gives an illuminating account of the rise of the third *Critique*'s problems in this context. Theodore Brown ("From Mechanism to Vitalism in Eighteenth Century English Physiology," *Journal of the History of Biology* 7, no. 2 (1974): 179–216) addresses vitalism in the context of British physiology, while vital forces applied especially to nervous system are in the focus of Stanley Jackson, "Force and Kindred Notions in 18th Century Neurophysiology and Medical Psychology," *Bulletin of the History of Medicine* 44 (1970): 539–54. About those forces and fluids stranger than usual physical forces, see also Rittersbuch, *Overtures to Biology*. Previously, Oswei Temkin tried to state the meaning of materialism and vitalism in those times ("Materialism in French and German Physiology of the Early Nineteenth Century," *Bulletin of the History of Medicine* 20 (1946): 322–30).

70. Richards, "Kant and Blumenbach on the *Bildungstrieb*: A Historical Misunderstanding."

71. Caneva, "Teleology with Regrets." Sloan indicated that, although Blumenbach seemed to accept the thesis about purposiveness, he nonetheless disregarded the Kantian perspective on natural history, insisting more and more on the "total habit" of animals in order to classify them, thus following a post-Linnean tradition that Kant opposed when he highlighted after Buffon the importance of historical criteria in natural history ("Buffon, German Biology, and the Historical Interpretation of Biological Species," 129–30).

72. *KU*, §81.

73. Hans Ingensiep, "Die biologischen Analogien und die erkenntnistheoretischen Alternativen in Kants *Kritik der reinen Vernunft* B §27," *Kant-Studien* 85 (1994): 381–93; Zöller, "Kant on the Generation of Metaphysical Knowledge"; Julius Wubnig, "The Epigenesis of Pure Reason," *Kant-Studien* 60 (1968): 147–52; and Arthur Genova, "Kant's Epigenesis of Pure Reason," *Kant-Studien* 65 (1974): 259–73.

74. Zöller, "Kant on the Generation of Metaphysical Knowledge," 74, subscribes to this last option, while Hans Ingensiep insists on the rhetorical or expository character of this appeal to epigenesis.

75. *KRV*, A834/B862: the systems of knowledge have their "scheme," as a "germ, *ursprüngliche Keim*," in reason.

76. Ak. 18, 223.

77. On Kant's precritical writings, Frederick Beiser's introduction is precious (Beiser, "Kant's Intellectual Development, 1746–1781," in *Cambridge Companion to Kant*, ed. Paul Guyer, 26–61). Indications are given in Eric Watkins, *Kant's Conception of Causality* (Cambridge: Cambridge University Press, 2005), Kuehn, "Kant's masters," and Michel Puech, *Kant et la causalité* (Paris: Vrin, 1990), sec. 1, especially concerning the reappraisal of the more common-sensical theory of the

physical influx, against the Leibnizian preestablished harmony. On *Populärphiloso-phie*, see Zammito, *Kant, Herder and the Birth of Anthropology*, ch. 3.

78. On this point, see Watkins, ed., *Kant's Conception of Causality*.

79. Friedman, *Kant and the Exact Sciences* (Cambridge: Cambridge University Press, 2001) and Vittorio Mahieu, *L'opus postumum di Kant* (Naples: Bibliopolis, 1991).

80. Or, with Guyer's words: "it seems reasonably clear that he wanted to make a transition from pure philosophy to an empirically applicable physics by constructing a system of forces that could ultimately be seen as deriving from some single elementary force, and that he wanted also to make a division of the fundamental types of *matter*. In the later division, the distinction between organic and inorganic matter is always primary, and Kant often presents the purposiveness evident in the former but not in the latter as the basis of this distinction." ("Organisms and the Unity of Science," in *Kant and the Sciences*, ed. Eric Watkins, 277).

81. Those critics are well expressed in his letter to Tieftrunck, April 5, 1798, Ak. 13.

82. For a comprehensive interpretation of this, see Mahieu, *L'opus postumum di Kant*, 220.

83. Ak. 21, 113.

84. About the organisms in the *Opus postumum*, see Wolfgang Riese, "Sur la théorie de l'organisme dans l'*Opus postumum* de Kant," *Revue philosophique* 3 (1965): 326–33; and Paul Guyer, "Organisms and the Unity of Science," in *Kant and the Sciences*, ed. Eric Watkins, 270–80, which studies the transformations of the arguments that proved the irreducibility of organisms in the *Critique of Judgment*. The striking point is that here Kant seemed to orient himself toward an organistic philosophy of nature that was quite forbidden by the critical perspective. Riese concludes: "the critical character of the determination of organism seems often to be sacrificed in favour of some more categorical and somehow dogmatic formulas" (p. 329). Guyer stresses the fact that arguments in the *Opus postumum*, like the one that is supposed to prove the immateriality of an organizing principle, are paralogisms according to the first *Critique* (p. 277).

85. About this: Claude Piché, "Kant et les organismes non vivans," in *La nature*, ed. L. Cournaraie and P. Dupond (Paris: Ellipses, 2001), 83–93; Huneman, *Métaphysique et biologie*, ch. 6; and Zammito, *The Genesis*.

86. Lenoir, "Göttingen School," *Strategy of Life*, and "Kant, Blumenbach and Vital Materialism in German Biology," *Isis* 71 (1980): 77–108.

87. Timothy Lenoir, "Generational Factors in the Origin of *Romantische Naturphilosophie*," *Journal of the History of Biology* 11, no. 1 (1978): 57–100.

88. James Larson, "Vital Forces: Regulative Principles or Constitutive Agents? A Strategy in German Physiology, 1786–1802," *Isis* 70 (1979): 235–49.

89. Jardine, *Scenes of Inquiry*, 36.

90. David M. Knight, *Science in the Romantic Era* (Aldershot: Ashgate, 1998); Nicholas Jardine, "*Naturphilosophie* and the Kingdoms of Nature," in *Cultures of Natural History*, ed. Nicholas Jardine et al. (Cambridge: Cambridge University Press, 1996), 230–45; and Isle Jahn, "On the Origin of Romantic Biology and Its Further Development at the University of Jena between 1790 and 1850," in *Romanticism in Science*, 75–89.

91. It should be mentioned that Kant's distinction, in his framework, is a kind of re-assessment of a classical metaphysical distinction from Leibniz, between logical principles imposed on all possible worlds and *"maximes de convenance"* (*Discours de métaphysique*, §3), which yield for God an imperative of purposiveness in the world he chose to create. (On Kant and Leibniz and the regulative principles, see Herbert J. Paton, "Kant and the Errors of Leibniz," in *Kant Studies Today*, ed. Lewis W. Beck (Lasalle, IL: Open Court, 1969), 301–21; and Anselm Model, *Metaphysik und reflektierende Urteilskraft, Untersuchungen zur Transformierung des leibnizschen Monadenbegriffs in der KU.* (Frankfurt: Athenaüm, 1987).)

92. On Geoffroy Saint-Hilaire's reference to embryology in order to establish homologies, see *Mémoire sur l'organisation des insectes*, pp. 75ff. Comments in Balan, *L'ordre et le temps*; and Dov Ospovat, *The Development of Darwin's Theory: Natural History, Natural Theology, and Natural Selection, 1838–1859* (Cambridge: Cambridge University Press, 1981). On some filiations of Kantian thought in Geoffroy's transcendantal anatomy, Philippe Huneman, "From Comparative Anatomy to the 'Adventures of Reason,'" *Studies in History and Philosophy of Biological and Biomedical Sciences*, 37, 4 (2006): 627–48. On German biologists' influence together with Geoffroy's influence on later morphology, Philip Rehbock, *The Philosophical Naturalists: Themes in Early Nineteenth-Century British Biology* (Madison: University of Wisconsin Press, 1983), ch. 1.

93. The idea of a unique animal type was intuited by Buffon (*Histoire naturelle*, vol. 4, pp. 135ff.) and then Diderot (*Pensées sur l'interprétation de la nature*, §12 (1753)), in a somehow evocative formulation. In the *KU* §80, Kant uses analogous formulations in order to finally deny any radical "archaeology of nature," which means a reconstitution of all the type species from a unique original form. Let's notice that in his chapter "Histoire naturelle de l'âne," Buffon also recognized the overwhelming impression that multiple species share a same type so as to, in the end, reject the hypothesis of transformation that would rely on such impression, since it lacks any empirical evidence. On those issues, see Huneman, "From Comparative Anatomy to the 'Adventures of Reason.'"

94. "Transcendantal Anatomy," in *Romanticism and the Sciences*, ed. Andrew Cunningham and Nick Jardine (Cambridge: Cambridge University Press, 1990), 144–60.

95. On Goethe's types, see Timothy Lenoir, "Morphotypes in Romantic Biology," in *Romanticism and the Sciences*, ed. Andrew Cunningham and Nick Jardine, 120–29, and "The Eternal Laws of Form: Morphotypes and the Conditions of Existence in Goethe's Biological Thought," in *Goethe and the Sciences: A Reappraisal*, ed. F. R. Amrine et al. (Dordrecht: Reidel, 1987); Robert J. Richards, *The Romantic Conception of Life: Science and Philosophy in the Age of Goethe* (Chicago: Chicago University Press, 2002), 413–94; Ron Brady, "Form and Cause in Goethe's Morphology," in *Goethe and the Sciences: A Reappraisal*; and Stéphane Schmitt, "Type et métamorphose dans la morphologie de Goethe, entre Classicisme et Romantisme," *Revue d'Histoire des Sciences* 54 (2001): 495–522. On the idea of type between Cuvier and Goethe, see Hans-Jorg Rheinberger, "Aspekte des Bedeutungswandels im Begriff organismischer Ähnlichkeit vom 18. zum 19. Jahrhundert," *History and Philosophy of the Life Sciences* 8 (1986): 242–47.

96.   The shifting sense of types between Goethe and Kant had been analyzed by Jardine, *Scenes of Inquiry*, 39; and Huneman, "From Comparative Anatomy."

97.   On Oken, see Jardine, *Scenes of Inquiry*, 45, and *"Naturphilosophie* and the Kingdoms of Nature," in *Cultures of Natural History*, ed. Nicholas Jardine et al. (Cambridge: Cambridge University Press, 1996), 230–45. Also Richards, *Romantic Conception*; and Stéphane Schmitt, *Histoire d'une question anatomique: la répétition des parties* (Paris: Editions du Museum d'Histoire Naturelle, 2005), and chapter 6 in this volume.

98.   Eveleen Richards provides some analyses of this trend in morphology, situated in its English political context and compared with the legacy of German romantic biology that deeply influenced it. ("'Metaphorical Mystifications': The Romantic Gestation of Nature in British Biology," in *Romanticism and the Sciences*, ed. Andrew Cunningham and Nick Jardine, 130–43.)

99.   On *Naturphilosophie* applied to life sciences, Richards, *Romantic Conception*, is now a main reference; on Schellling, see, among many papers and books, Judith Schlanger, *Schelling et la réalité finie* (Paris: Puf, 1966), and Andrew Bowie, *Schelling and Modern European Philosophy: An Introduction* (London: Routledge, 1993); on Hegel, see Frederick Beiser, *German Idealism: The Struggle against Subjectivity* (Chicago: University of Chicago Press, 2003). It is worth reminding that, as philosophical projects, Hegel's and Schelling's philosophies of nature differ completely, even though they share the same genealogy; a general framework for understanding philosophies of nature in their relation to Kant is suggested in Philippe Huneman, "From the *Critique of Judgment* to the Hermeneutics of Nature: Sketching the Fate of the Philosophy of Nature after Kant," *Continental Philosophy Review* (2006) 3: 1–34.

100.   On Owen, see Balan, *L'ordre et le temps*; Phillip Sloan, introduction to *The Hunterian Lectures in Comparative Anatomy, May and June 1837*, by Richard Owen, ed. Phillip Sloan (Chicago: University of Chicago Press, 1992). On Owen and Oken, see Olaf Breidbach and Mihael Ghiselin, "Lorenz Oken and *Naturphilosophie* in Jena, Paris, and London," *History and Philosophy of the Life Sciences* 24 (2002): 219–47. More precisely on Owen and archetypes in a German fashion, see Phillip Sloan, "Whewell's Philosophy of Discovery and the Archetype of the Vertebrate Skeleton: The Role of German Philosophy of Science in Richard Owen's Biology," *Annals of Science* 60 (2003): 39–61. Stephen Jacyna highlighted a quite unnoticed consequence of *Naturphilosophie* or "romantic" biology on modern biology, with cell theory. The idea of an original element of life—as is the cell—was provided in the intellectual atmosphere by this romantic biology and its leitmotiv of the invisible original type, whereas such an atmosphere was strongly rejected by the empiricist promoters of this theory. ("Romantic Thought and the Origins of Cell Theory," in *Romanticism and the Sciences*, ed. Andrew Cunningham and Nick Jardine, 161–68, and "The Romantic Program and the Reception of Cell Theory in Britain," *Journal of the History of Biology* 17, no. 1 (1984): 13–48.)

101.   Jardine, *Scenes of Inquiry*, 50.

102.   One example of this misunderstanding alteration of Kant's philosophy into an organicism is given by Fries; see Frederick Gregory, "'Nature is an organized whole': Fries's Reformulation of Kant's Philosophy of Organism," in *Romanticism in Science: Science in Europe, 1790–1840*, 91–101. On Reil, see Richards, *Romantic*

Conception, ch. 7. On Kielmayer and *Naturphilosophie*, see Thomas Bach, "Kielmayer als 'Vater der Naturphilosophie'? Anmerkungen zu seiner Rezeption im deutschen Idealismus." In *Philosophie des Organischen in der Goethezeit: Studien zur Werk und Wirkung des Naturforschers Carl Friedrich Kielmayer (1765–1844)*, ed. Kai Torsten Kanz (Stuttgart: Steiner, 1994), 232–51; and Richards, *Romantic Conception*, ch. 6.

103. Peter Hans Reill, "Vitalizing Nature and Naturalizing the Humanities in the Late Eighteenth Century," *Studies in Eighteenth Century Culture* 28, ed. J. C. Hayes and Ted Erwin (Baltimore: John Hopkins University Press, 1999), 361–38. He recently expanded his views on Enlightenment vitalism in the book *Vitalizing Nature in the Enlightenment* (Chicago: Chicago University Press, 2005).

104. A recent and detailed account of the rise of *Naturphilosophie* in the context of the reception of critical philosophy and dissatisfaction regarding it is Beiser, *German Idealism*.

105. Jardine, *Scenes of Inquiry*; Knight, *Science in the Romantic Era*; and Richards, *Romantic Conception*.

106. Richards, *Romantic Conception*, ch. 14.

107. Sloan, "Buffon, German Biology, and the Historical Interpretation of Biological Species" and "Preforming the Categories."

108. "Strange" if we consider that Von Baer was the other great biologist of the day.

109. Ospovat, *The Development of Darwin's Theory*, ch. 5, 115ff.

110. Coleman, *Biology in the 19th Century*, stresses the pervasive influence of a historical model of explanation in the life sciences through the nineteenth century, whether or not authors are dealing with transformism. With some other studies, such as Pietro Corsi, *The Age of Lamarck: Evolutionary Theories in France, 1790–1830* (Berkeley: University of California Press, 1988), he renders quite irrelevant the opposition transformism/fixism to apprehend biology in this period. It is least of all appropriated when German post-Kantian biology comes to the point.

111. On the law of recapitulation in Kielmayer, see Richards, *Romantic Conception*, 246–48; chapter 5 in this volume; and William Coleman, "Limits of the Recapitulation Theory: Carl Friedrich Kielmayer's Critique of the Presumed Parallelism of Earth History, Ontogeny, and the Present Order of Organisms," *Isis* (1973): 341–50. On the law itself, see Stephen Jay Gould, *Ontogeny and Philogeny* (Cambridge: Belknap Press, 1977); Peter Bowler, *Evolution: The History of an Idea* (Berkeley: University of California Press, 1984); for a critique of those views, emphasizing a clear commitment to progressism shared even by Darwin, see Robert J. Richards, *The Meaning of Evolution: The Morphological Construction and Ideological Reconstruction of Darwin's Theory* (Chicago: Chicago University Press, 1992).

112. *Entwicklungsgeschichte der Thiere*, Scholius V.

113. George Canguilhem et al., *Du développement à l'évolution* (Paris: Puf, 1962); and Kai Torsten Kanz, *Philosophie des Organischen in der Goethezeit* (Stuttgart: Franz Steiner Verlag, 1994).

114. Johann G. Herder, *Ideen zur Philosophie der Geschichte der Menschheit* [1784], éd. B. Suphan (reprint, Hildesheim: G. Olms Verlag, 1967), Book 10, ch. 1.

115. Some classic papers on the troubled meaning of transformism by German authors in the first half of nineteenth century, such as those by Kant, Schopenhauer,

and Herder, are to be found in *Forerunners of Darwin, 1745–1859*, ed. B. Glass et al. (Baltimore: Johns Hopkins University Press, 1959). Herder and Schopenhauer appear there as real tranformists, unlike Kant.

116.   Richards, *Meaning of Evolution*, gave already a detailed demonstration of the importance of the progressive *Naturphilosophische* scheme of evolution for Charles Darwin, arguing that both of them subscribed to a form of recapitulation.

117.   For this classical interpretation, see Bowler, *Evolution*; and Michael Ruse, *The Darwinian Revolution* (Chicago: University of Chicago Press, 1979).

I

Pre-Kantian Revival of Epigenesis

Caspar Friedrich Wolff's
De formatione intestinorum (1768–69)

Jean-Claude Dupont

Abstract

In 1768, the German embryologist Caspar Friedrich Wolff (1733–94), recognizing the uselessness of further theoretical argument with Bonnet and Haller on invisibility, felt at an impasse in advancing his views about *vis essentialis* and formative causes, though he would return to these later in his discussions with Blumenbach. For the moment, it was necessary to produce a work whose professionalism would equal that of Haller on the heart: hence he produced *De formatione intestinorum*. After summarizing ideas about generation developed in his medical thesis of 1759 (*Theoria generationis*), Wolff minutely analyzed the development of the internal organs, mainly the digestive system. Recounting the genesis of this work, my chapter suggests that—because it described intermediate embryonic forms, and because it broke with the dominant preformationism as well as with the ancient epigenesis inherited from Aristotle—it appears to be the first great text of modern embryology. Setting the *vis essentialis* debate apart from the epigenetic problem in embryology, this text made realization of the Kantian teleomechanistic research program an actual possibility.

Introduction

According to the German embryologist Caspar Friedrich Wolff (1733–94), the body cannot be conceived without the existence of a epigenetic force. Wolff thus introduces a deep transformation of biology, releasing it from a conception of organization strictly modeled after the physical world. Even though this liberation is still only partial, and even though it is to Blumenbach and not to Wolff that Kant refers for that reason in the third *Critique*,

Wolff's embryological works do represent a condition of realizability of the Kantian project for the biology.

However, this required first a scientific demonstration of the reality of the epigenesis. This actual demonstration came in the *De formatione intestinorum*, published in the Reports of the Academy of Sciences of Saint Petersburg.[1] Once he had reminded readers of the ideas about generation that he had already explained in his doctoral thesis in medicine in 1759,[2] Wolff went on to study the development of the internal organs, mainly those of the digestive system, in minute detail. Because it described intermediate embryonic forms and because it broke with the prevailing preformationism and the ancient theory of epigenesis inherited from Aristotle, *De formatione intestinorum* appears to have been the first great work of modern embryology. What I propose here is not an analysis of this very technical work, but rather an understanding of its genesis, which can open the way to a better understanding of the specificity and peculiarity of Wolff's biology, situated as it is between the mechanistic biological tradition and the new German embryology. This should highlight the character of novelty proper to Wolff's work, compared with the previous vindications of epigeneticism, and the awareness he had of proposing an innovating framework for further empirical inquiry. Thus, he was a major reference in this new epistemological situation that Kant had to face when he turned to elaborate his own metaphysical assessment of an emergent biological science.

1. *Theoria generationis* (1759)

Wolff's complete works include the *Theoria generationis*, his 1759 doctoral thesis in medicine, along with the subsequent 1764 and 1774 versions (which are far from being simple replications of the initial text), and approximately thirty monographs published in the reports of the Academy of Sciences of Saint Petersburg, among which is included *De formatione intestinorum*. His unpublished manuscripts, his correspondence with Haller, and his unfinished treatise on monsters may be added to these.[3]

One may say that Wolff's work arose from a dispute with embryology as dominated by preformation theory. To understand this, and to understand how Wolff came to describe in great detail how chicken intestines were formed in his *De formatione*, which would become his most influential publication, it is necessary to recall his *Theoria generationis*.

Caspar Friedrich Wolff was the pupil of the famous person of the same name, Christian Wolff, who himself was a follower of Descartes and Leibniz and who influenced Kant, which undoubtedly led to the acquisition of the habit of providing preliminary definitions and adopting a rigorous terminology.

"Let us consider," says Wolff, "as a principle of generation that capacity of the body which allows this formation. And the mode in accordance with which it acts constitutes the laws of the generation called for by the Very Illustrious Haller. Those who develop systems involving preexistence do not explain generation, but are content to assert that it does not occur. And I could in no way consider that those who have not explained any part, any attribute of the body on the sole basis of the principles provided by the theory, have succeeded in explaining generation. They have only engaged in discourses on the matter, sometimes learned, true, and elegant discourses."[4]

If in the definition and exposition of the method of the project of a Theory of Generation, Wolff at once refers to "the Very Illustrious Haller," this is to distinguish himself from Albrecht von Haller from the outset. Wolff's intention comes clearly to the fore: to reappraise embryology as a historic and descriptive empirical science. Against these "sometimes learned, true and elegant discourses" Wolff sets the true principles of a "rational anatomy." It is his ambition to develop an *anatomia rationalis* based on a Theory of Generation. While he is interested in consequences, shape, and processes that preserve the shape of the body, in other words in anatomy and physiology, this is to discover better the deepest causes, the processes modifying or creating it—that is, the laws of generation.

The precise modalities of this generation, which he defined as nutrition and vegetation, are described in the three parts of Wolff's work. So, the first part of *Theoria generationis* deals with plants, which must be studied first because they are easier to know than animals. The second part deals with the generation of animals. Based on the findings of the two first parts, the third part deals more generally with the laws of generation of the organic bodies.

Fluids are constantly absorbed from the ground into the plant, distributed throughout it, and evaporated through the leaves. Some force must cause this movement. "It must be presumed to exist, if we admit plants and their nutritive liquid, something which has been confirmed through experimentation. This suffices for our commentary and I shall call it "the essential force of plants (*vis essentialis*)."[5] Young plants (and young animals) arise from raw material, an inorganic, amorphous, and undifferentiated substance. Within, vesicles (*vesiculae*) or corpuscles (*globuli*) appear like small spherical cavities filled with liquid. On the other hand, vessels (*vasae*) appear. It is the stream of the nutritive fluid that proves decisive in the formation of tissues. Flowing slowly, the fluid loses its volatility, becomes sticky, and creates vesicles; flowing rapidly (in the same direction), it creates vessels. Vesicles then become organs that differentiate and appear one after the other. The streaming of liquid and the tendency to coagulate, which is in fact a force of cohesion, are sufficient to explain the structure of plants and animals without any need of preexistent structures.

Development is thus based on two factors: the essential force and the tendency of plant and animal fluids to solidify. In plants there is a "vegetative point" at the end of every growing stalk out of which leaves, flowers, and fruits develop by secretion and solidification. The particular type of structure produced at a given time depends on the level of the nutritive liquid reaching the vegetative point. Vegetal structures are thus "leaves" essentially modified by particular circumstances. One can, moreover, peel away leaves or buds at the end of their growing stalks and find them in a smaller, rudimentary form beneath. The new leaves that one finds enfolded in miniature inside the other older ones establish not preexistence but rather epigenetic development.

Animal development is treated in a similar way. The "germ" is an unorganized, amorphous substance secreted by the genital organs of the parents. It becomes organized gradually only after conception, which is a sort of nutrition: male semen is a nutritive substance, a food that reaches the seed. Wolff describes the heart, which appears in the shape of a tube, not pulsatile, and not connected to arteries and veins. Material has to enter the embryo from the yolk, and "it follows that the nutritive particles pass from the egg to the embryo and that there is a force by which this is realized. In similar fashion, I would call it essential force."[6] As in plants, *vis essentialis* indeed always accompanies the tendency to coagulation (*solidescibilitas*) in animals. The process of secretion and solidification begins in the seed of the plant. In animals it begins in the yolk. It acts in an orderly manner, each part secreting the other after its own formation. Wolff particularly describes the moment preceding the appearance of the pulsations of the heart. In the vascular area, the substance of the embryonic disc forms islands that fill with blood. These islands precede the formation of the heart and the vessels, which form themselves by thickening the walls.

Let us note the process of auto-formation. The first parts formed produce the others, which form and extend them. During incubation heat facilitates the dissolution of the yolk, which can be assimilated by the embryo. But minute observation of the germinal spot before heat begins to take effect shows that "neither heart, nor vessels, nor traces of red blood" are perceptible. The spot is a mere imprint, or trace, of small vesicles, or of small, more or less agglutinated, spheres (*globuli*). Only the observation of the movements of fluids in the vascular area gives the impression of the preexistence of vessels. Wolff was certain that even the best microscopes could not reveal more of this: "It is a fairy tale belief to think that constituents can remain hidden because of their infinitely small size and become visible subsequently."[7] Contrary to Malpighi's assertions, hen eggs contain nothing resembling the future bird.

In his foreword, Wolff states the intention of the last part: "I now give the general laws of the generation of organic bodies as derived, on the one hand, from the previous parts, and on the other hand also, to make them more universal, I give them as demonstrated from the other principles."[8] With his model of secretion and solidification and *vis essentialis*, Wolff thought he was able to offer a complete explanation of development. "With the capacity of the nourishing fluid to solidify, essential force establishes the sufficient principle of all vegetation [development], both for plants and for animals."[9] It need not postulate the existence of preformed parts, or representative molecules in the embryo. Development is a modification of structures, a kind of construction effectuated by natural forces. The various parts of the embryo successively emerge from an unorganized substance in simple shapes completely different from the shape that these parts will take on afterward.

2. Criticisms of *Theoria generationis*

Wolff's rational anatomy was intended to compete directly with the embryology of the great Albrecht von Haller. It was indeed about one year after the publication of his great work on the formation of the heart[10] that Haller received *Theoria generationis* from Wolff, perhaps sent in the hope that Haller would abandon his defense of the thesis of the preexistence. The exact opposite happened. Between 1759 and 1777 (the year of Haller's death) one of the most famous controversies in the history of sciences ensued.[11] This controversy over embryology and philosophy survives in a series of publications: the correspondence between Wolff and Haller from 1759 to 1777 (with nine letters written by Wolff);[12] Haller's reviews of each of Wolff's works, the *Theorie von der Generation* (1764),[13] the German translation of Theoria generationis considerably revised and expanded in response to Haller; the eighth volume of Haller's *Elementa physiologiae corporis humanis* (1766);[14] and the revised version of the treatise *On the Formation of the Heart in the Chicken* (1767).[15]

Haller praises the quality of Wolff's observations in *Theoria generationis* but disputes its theoretical basis. His opposition is mainly and classically represented by theoretical objections regarding the transparency and the solidarity of organs, and the nature and indefiniteness of the *vis essentialis*. Haller's most traditional argument appeals to the invisibility of preexistent parts. Although parts are invisible, we must not deduce that they do not exist because incubation only makes visible what was originally invisible. Preexistent structures are transparent, semi-fluid. Blood vessels are "still hidden," transparent, soft, or flattened. The initial mass nevertheless conceals the structure of the future animal. There is even experimental support for this: if one pours the white of the egg into alcohol (wine spirits), which is not *vis essentialis*, vessels and internal organs become visible. Wolff has made the mistake of assimilating the invisible to the nonexistent.

The second argument is represented by Haller's physiological objection, namely, the impossibility of united organs taking form successively. Harvey's old theory of epigenesis conceived development as the adding of organs that are formed successively but reach their definitive type at once. This epigenesis without transitional embryonic structures "sometimes seemed to be a sort of split up preformationism."[16] The preexistence of organs makes them at once united. Even though one imagines differentiation in the growth of parts, organ solidarity is maintained. The fragmented conception of the body that epigenesis actually conceived of as the successive creation of definitive, isolated organs created the problem of the subsequent establishment of a solidarity, without which the body could not coordinate the functioning of its parts. The old theory of epigenesis partially maintained certain presuppositions concerning preexistence, such as the absence of transitional forms, while creating the new problem of the solidarity of new organs.

Charles Bonnet would sum up the last of the anti-epigenetic arguments, which concerns *vis essentialis*, in this way:

> A force is always in itself indefinite: it can just as well produce one particular effect as another. One requires something preexistent determining this force to produce a certain effect rather than another that it could also produce. But if there is nothing preformed in the matter that essential force organizes, how will this force determine the production of an Animal, rather than a Plant, and a certain Animal in preference to another one? Why will essential force produce a certain organ in a certain place and not another one? Why will this organ constantly have the same shape, the same proportions, and the same situation in a given genus? Why? But this would give rise to an infinite number of whys, and there would not be any use resorting to the matter on which essential force works because this matter is as indifferent to one shape as it is to another, as force itself is to one modification or another. Besides, what I have just said regarding the indefiniteness of this essential force that our epigenesist would like to introduce into Physics applies to the organic molecules that another epigenesist considered able to produce a new Nature by themselves.[17]

3. *De formatione intestinorum* (1768–69)

The objections made by the partisans of preexistence were also based on data from recent embryological observations. Haller's treatise of 1758 on the formation of the heart was already the outcome of considerable work. Following Wolff's objections, notably in *Theorie von der Generation*, Haller would again make his observations and again engage in experimentation. This would lead him to confirm the positions concerning three main embryological points: the formation of vessels, the formation of the heart, and the continuity of membranes, positions that he would essentially develop in *Elementa physiologiae and Opera minora*.

We now understand Wolff's deep-seated motives. He is aware of the impasse that yet another discussion of *vis essentialis* (and formative causes, problems he will bring up again in his discussions with Blumenbach) would represent, as well as the uselessness of engaging in theoretical argumentation with Bonnet and Haller on invisibility. To ground his theory, it was necessary to produce a work whose professional skill would equal that of Haller on the heart: this would be the *De formatione intestinorum*.

I shall not summarize all of Wolff's observations concerning the envelopes of the embryo and the digestive system here but shall emphasize that the full originality of Wolff's theory of epigenesis undoubtedly appears better in *De formatione* than in *Theoria generationis*. Wolff proposes a model of differentiation that cannot be assimilated to that of Aristotle and Harvey. If he rectifies the old theory of epigenesis, it is by suggesting that "formation takes place by a series of different creations of systems (nervous, muscular, digestive, etc.), rather than of isolated organs, each of them forming a relatively autonomous whole."[18] The epigenesis in question is simultaneously that of connective systems, and not of organs. At that point, Haller's objection concerning the solidarity of organs is no longer of any concern to Wolff. So Wolff's treatise is innovative in this regard.

The other key concept is that of intermediate embryonic forms, of rough shapes. The intestinal layers that will form the walls of the intestinal tube merge and form a cavity to produce all the parts of the intestinal tube. A space appears between the two layers of the *sutura* to form the opening of the canal. "So I ask," says Wolff, "are these layers the complete intestine? Certainly, nobody will maintain that. I, therefore, conclude from this that complete, formed parts did not always exist, but that they were formed at the proper moment after conception."[19]

But Wolff still concludes:

> I think that this is a major argument in favor of epigenesis. We can certainly conclude from it that parts of the body did not always exist, but were produced successively and, moreover, something of the way in which this production takes place. Since I am not saying that these parts are produced by a meeting of particles, either through the mode of fermentation, or through mechanical reasons and causes, I purely and simply say "produced." If one considers these parts at a little more mature stage, they supply a basis for a new argument. The rough shapes are now there, but ready in such a way that one easily recognizes that they are not complete, instantly formed parts, but rather rough shapes of that sort to be transformed into parts of that kind.[20]

It would obviously be naïve to believe that with *De formatione* Wolff's empirical observations brought proof of the epigenetic development of the embryo in any particularly definitive way, thus abruptly reducing to nothingness the relentless opposition he faced from his influential colleagues.

The difference between Haller and Wolff lies mainly in the interpretation of the invisible. Haller would make two reports on *De formatione*, one in 1770 and one in 1771, in which he still took up the defense of the theory of transparency.[21] Debate about embryology ceased, however, as is evident in the dropping off of the correspondence between Wolff and Haller (they exchanged only two letters in the years leading up to 1777, the year of Haller's death). As for Wolff, he would defend the theory of epigenesis until his own death. He would pursue his research on the heart and the vessels, on twin forms, and on monsters, which he published in the reports of the Academy of Saint Petersburg. His teratological works would clarify the epigenesis of abnormal forms of development. Monsters were the product of the usual laws of the generation, when by chance they operated under unusual conditions. In a 1783 letter to Johann Albrecht Euler, Wolff refuted preexistence in Bonnet's sense, notably the proof by the continuity of the membranes:

> So, all that is in these notes against epigenesis is reduced to the continuation of one of the skins of the yolk (of the insides that form during incubation) in the embryo, to which I replied two things twenty years ago: 1) that it does not exist as he said and as evolution would seem to prove; 2) that if it existed, as the late Mr. de Haller wanted, that would not prove anything. The robust refutation of my objections by the late Mr. de Haller consists of his having believed that he observed in a certain stage of incubation (it cannot have been at the beginning of the third day) the vessels of the umbilical area already completely formed and existing. Mr. de Haller (in this letter to Mr. Bonnet) adds a single word of mockery. He believes, he says, that after that I should stop worrying. But I wrote my two essays on the formation of the stomach and the intestines after that. Mr. de Haller called them very important, and he has never responded to them. So it seems so that Mr. De Haller himself has ceased to worry.[22]

4. On the Special Essential Power (1789)

Wolff's embryology responds to the arguments concerning the transparency and the solidarity of organs, but the theoretical problem of the vis essentialis still remains. It is this problem that makes for a better understanding of Wolff's biology, situated as it is between the Cartesian and the Kantian biological tradition.

Aiming to avoid repetition, Wolff claims to have read works concerning epigenesis at the time of the writing of the third part of the *Theoria generationis*. Trying to describe his *vis essentialis*, he especially discusses the thesis of John Tuberville Needham and Christian Gottlieb Ludwig.[23] In his understanding of life, Wolff distances himself clearly from "mechanistic medicine."[24] It was Wolff's ambition, as it was Descartes,' to understand the origination of functions or vital phenomena. But is development (*evolutio*) a product of mechanistic laws? Are these laws insufficient to build a body? It was the following questions, central in the wake of Descartes' work, that "rational anatomy" would ultimately have to answer: "How are life and 'mechanics' combined together in natural organic bodies? Are they both dependent on a common cause, or does one of them cause the other? And if the latter is true, what does life bring to the machine, and vice-versa?"[25]

Wolff would answer by distancing himself clearly from mechanistic explanations of vital phenomena and, curiously, by drawing closer to Georg-Ernest Stahl. However, this link must be considered with caution. Wolff would have inverted several of the relationships established by the theory of preexistence. Generation was accordingly conceived of as simple growth. For Wolff, it was growth that was thought of as a sort of generation (itself defined as a nutrition), reproduction of each part on a reduced scale, presided over by the process of secretion and solidification and essential force. According to the partisans of preexistence, the parts of the vegetable being likened to a complete and autonomous being (twigs are small, quite finished trees growing on the trunk of another tree), the animal was a model for the vegetable (contrary to appearances). According to Wolff, however, the vegetable becomes a model for the animal, as far as there are in the animal purely vegetative forces shielded from the power of the soul. Comparing animals and plants from that perspective freed Wolff of the limitations of mechanistic causality, but without making him fall into animism.

Actually, according to Malpighi (often quoted by Wolff), the study of lower forms of life simply had to clarify the morphology of the higher. By constantly underscoring the plant-animal analogy, Wolff not only set out, as did his predecessors (Malpighi, for example), to follow an old tradition but pursued a deeper theoretical aim that went beyond Descartes and Stahl.

Wolff was familiar with mechanistic explanations, as well as with Cartesian physiology and embryology. A full-grown body can behave mechanically as a machine, and in affirming that Wolff followed faithfully in Descartes' footsteps. Haller, moreover, viewed Wolff's system as being mechanistic. But his *vis essentialis* introduced a serious difference with regard to the mechanism involved. Wolff advocated epigenesis, but placed that successive construction of organic structures under the control of the *vis essentialis*. While for the theory of preexistence, to which Cartesian philosophy could thus indeed lead, growth could simply be explained by nutrition, the ongoing production of new organs and new forms of interaction among organs indeed seemed to require something else, that is, a *force*. The body did not remain a machine throughout the entire course of it development. It became a machine only when it was full-grown.

Now Bonnet's and Haller's classic theoretical argument against *vis essentialis* is its indefiniteness. A hen egg becomes a hen; a peacock egg becomes a peacock. How could the same *vis essentialis* create different animals from the same unstructured mass? Haller rightly considered essential force as "blind," similar to Needham's *vires*, and generally speaking Wolff reproduced the theoretical weaknesses of the partisans of "imaginary" forces. Unlike Wolff, Haller was still very attached to traditional mechanism and altogether rejected confused notions of *vis essentialis* (Wolff), of *Lebenskraft* (F. K. Medicus), of *vegetative force* (Needham), and even of *forces attractives* (Buffon).

The nature of this force is actually difficult to clarify. "His *vis essentialis* is an illegitimate entity, and maybe the least original idea of *Theoria*. On the subject, it will be enough to note here that while Wolff distinguishes it from all the known physical forces, he does not make of it a simple regulating or final cause which would assure a constant specific term in the ontogeny."[26] To be able to build a body in a causalistic and mechanistic way, *evolutio* requires a purpose and, in addition, a sequence. Wolff's *vis essentialis* initially seemed to represent the alliance of two Aristotelian causes: the *causa efficiens* and the *causa formalis*. But Wolff rejected the finalism, and that means affirming that the *causa formalis* can no longer be a true *causa formalis*, as would be well shown in Wolff's subsequent opposition to Blumenbach.

In addition, the philosopher Christian Wolff agreed with Stahl for whom the vegetative process could not be reduced to a mechanism. It was seen above that his pupil Caspar Friedrich Wolff concurred with Stahl regarding a generative force in nature. But for Wolff, there was no animating, determining, steering vital movement. Within living substances, the *vis essentialis* acted as a physical force, a natural drive that nourished and made grow, and must not be considered a soul. Wolff's theory of epigenesis would eliminate the soul from earlier theories of epigenesis, but the theoretical consequences of the *vis essentialis* were not totally assumed.

Wolff found himself at a loss when faced with a tricky, perhaps unsolvable, problem. On his initiative, in 1782, the Saint Petersburg Academy of Science announced a competition for a prize on the question of the nature of the feeding powers causing the movement of plant juices. Johann Friedrich Blumenbach, Carl Friedrich von Born, and Wolff's own contributions were printed together in one manual in 1789.[27] Blumenbach and Born postulated a particular vital agent for the matter of organic bodies. And Wolff agreed with this. According to Wolff, this force determining the growth of living bodies was neither an *anima*, as Stahl thought, nor a *nisus formativus* or *Bildungstrieb*, as Blumenbach thought. It was by nature close, but not identical, to Newton's attractive force, a "special kind of attractive and repulsive force."[28]

This essential force depended not on the entire living substance but on one constitutive part. The plant substance attracted the homologous substance and replaced the heterologous substance. The essential power operative was a nourishing force (and the generation was vegetation and nutrition). The effects of this force remained simple ones of the attraction or aversion type. They could not themselves impose the shape of the body. The *vis essentialis* was driven by the qualities or the properties of the matter that are passed on from generation to generation (like the tendency toward solidification, for example).

The specific effects of *vis essentialis* on living entities were connected with circumstances and uncountable convergent causes. It was not a formative force. There was no presupposition of an inherent organization of the organism in Wolff's theory of epigenesis, but only a predisposition to future organization. He even rejected Leibniz's preestablished harmony as the philosophical source of the idea of preexistence. While Wolff accepted the idea of mechanical forces or powers as the source of life processes in general, and specific to them, there was no idea of an immanent teleological power. With Blumenbach, the source of embryonic organization would be admitted into the generative material and the organization would be accepted as a teleological fact. With this difference, Wolff remained very close to Blumenbach.

After Wolff, empirical investigations into the mechanisms of this new "generic preformation" (Kant) would become possible.[29] Leaving aside speculative analyses of the formative cause and the problem of the source of organization, the nineteenth century would rectify Wolff's theory of epigenesis as concerned essentially with embryological facts, like the question of embryonic layers, and then reinterpret the theory within the framework of cellular theory. In fact, the theory of preexistence had to remain until Louis Tredern, Christian Pander, and Karl Ernest von Baer opened new perspectives for embryology. For example, Wolff described the succession of what he considers the three layers enveloping the embryo—the pellucid area, the vascular area, and the false amnion—from which true amnion is released. But these enveloping layers still only constitute the successive arenas of the development of the embryo without really constituting it. In other words, even though the formation of the intestine results from the false amnion, the genetic link between the embryo and its envelopes was not yet established. With Pander, the blastoderm would become a germinal membrane out of which enveloping layers would develop as the result of the extra-embryonic development of leaves. Additional light would thus be thrown on the formation of the blastoderm.

But Wolff's contribution also lies in something more fundamental. The three questions of the transparency and the solidarity of organs, and of *vis essentialis* were closely connected, and that link leads us to a better understanding of the double denial of the preexistence of seeds and mechanical epigenesis and more especially the epistemological rupture operated by Wolff. According to him, the seriousness of a theory of generation consisted in the arrangement of what was actually seen, and not in what was supposed to exist. Preexistent organs could not exist because transitory embryonic structures followed one another. Essential force was an intellectual requirement deduced from this observation and not a physiological necessity conceived a priori. This essential force caused the succession of embryonic forms, the combination of the parts of the body; it was the explanation that made construction possible. It was specific to living entities and remained indefinite in nature; neither exactly physical (Buffon) nor mental (Maupertuis). Far from invalidating his thesis, this indefiniteness and this specificity guaranteed the status of embryology as an autonomous science, casting off any outside physical model of comprehensibility. In this sense, and because of his *De formatione intestinorum*, Wolff would be rightly celebrated as the creator of a new embryological science.

Notes

1.   "De formatione intestinorum praecipue, tum et de amnio spurio, aliisque partibus embryonis gallinacei, nondum visis." *Novi commentarii academiae scientiarum imperialis petropolitanae*, 12 (1768): 403–507; 13 (1769): 478–530. German translation by Johan Friedrich Meckel, *Über die Bildung des Darmkanals im bebrüteten Hühnchen* (Halle: Renger, 1812). French translation by Michel Perrin, *Caspar Friedrich Wolff, la formation des intestins (1768–1769)*, with introduction and notes by Jean-Claude Dupont (Turnhout: Brepols, 2003).

2.   *Theoria generationis* (Halle: Hendel, 1759; reprint, Hildesheim: G. Olms, 1966). German translation by P. Samassa, Ostwalds Klassiker der Exakten Wissenschaften no. 84–85 (Leipzig: Wilhelm Engelmann, 1896); Russian translation by A. E. Gaissinovitch and E. N. Pavlovski (Moscow: Edition of the Academy of Sciences of the USSR, 1950).

3.   For a bibliography, see Shirley Roe, *Matter, Life, and Generation: Eighteenth-Century Embryology and the Haller-Wolff Debate* (Cambridge: Cambridge University Press, 1981).

4.   *Theoria generationis*, 5.

5.   Ibid., §4.

6.   Ibid., §168.

7.   Ibid., §166.

8.   Ibid., 9.

9.   Ibid., §242.

10.   Albrecht von Haller, *Sur la formation du coeur dans le poulet; sur l'oeil, sur la structure du jaune, etc.*, 2 vols. (Lausanne: M. M. Bousquet, 1758).

11.   See Roe, *Matter, Life, and Generation*.

12.   See ibid.

13.   *Theorie von der Generation in zwo Abhandlungen erklärt und bewiesen* (Berlin: Friedrich Wilhelm Birnstiel, 1764; reprint, Hildesheim: G. Olms, 1966).

14.   *Elementa physiologiae corporis humani*, vol. 7 (Lausannae: M. M. Bousquet, S. d'Arnay, F. Grasset, Société typographique, 1766).

15. *Commentarius de formatione cordis in ovo incubato*, in *Opera minora*, vol. 2 (1767), ed. Maria Teresa Monti, Studia Halleriana VI (Basel: Schwabe, 2000), 54–421.

16. George Canguilhem et al., *Du développement à l'évolution au XIXe siècle* (Paris: PUF, 1962), 7.

17. Charles Bonnet, *Considérations sur les corps organisés, ou l'on traite de leur origine, de leur développement, de leur reproduction, etc.* (Amsterdam: M. M. Rey, 1762; 2nd ed. 1768; 3rd ed. 1779), vol. 22 of *Oeuvres d'histoire naturelle et de philosophie* (Neuchâtel: Fauche, 1779–83; reprint, Paris: Fayard, 1985), 467–68.

18. Canguilhem et al., *Du développement*, 8–9.

19. Wolff, *De formatione intestinorum*, §155.

20. Ibid.

21. Albrecht von Haller, "Review of *Novi commentarii academiae scientiarum imperialis petropolitanae*, including *De formatione intestinorum* by Wolff," *Göttingische Anziegen von gelehrten Sachen* (1770): 377–81; (1771): 414–16.

22. *Lettres à Mr. l'abbé Spallanzani de Charles Bonnet*, ed. C. Castellani (Milan: Episteme Editrice, 1971), 513.

23. Wolff, *Theoria generationis*, §231–35.

24. Ibid., §255.

25. Ibid., 9.

26. Canguilhem et al., *Du développement*, 9–10.

27. Johann Friedrich Blumenbach and Carl Born, *Zwei Abhandlungen über die Nutritionskraft welche von der Kayserlichen Academie der Wissenschaften in Saint Petersburg den Preis gethheilt erhalten haben. Nebst einer fernern Erlaüterung eben derselben Materie von Caspar Friedrich Wolff* (St. Petersburg: Kayserlichen Academie der Wissenschaften, 1789); Caspar Friedrich Wolff, *Von der eigenthümlichen und wesentlichen Kraft der vegetabilischen, sowohl als auch der animalischen Substanz* (St Petersburg: Kayserliche Academie der Wissenschaften, 1789).

28. On the debate with Blumenbach, see Rupp-Eisenreich, "Wolff, Caspar Friedrich," in *Dictionnaire du darwinisme et de l'évolution*, ed. Patrick Tort (Paris: PUF, 1996), vol. 3, pp. 4666–71; François Duchesneau, *La physiologie des lumières. Empirisme, modèles et théories* (Den Haagen and Boston: London Martinus Nijhoff, 1982), chap. 8; François Duchesneau, "Epigénèse et évolution: prémisses historiques," *Annales d'Histoire et de Philosophie du vivant* 6 (2002): 177–203.

29. Kant, *Kritik der Urteilskraft*, §81.

2

Kant's Persistent Ambivalence toward Epigenesis, 1764–90

John H. Zammito

Abstract

At B167 of the *Critique of Pure Reason* (1787), Immanuel Kant drew a remarkable analogy between the idea of epigenesis in the theory of generation and his own idea of transcendental philosophy. Not only does this analogy still demand convincing elucidation, it raises questions about exactly what Kant understood by epigenesis and how he felt about that idea. My contention will be that neither before nor after this now somewhat famous analogy was Kant entirely comfortable with the idea. Indeed, I argue that Kant proved resolutely hostile to the idea in both published and unpublished sources from his first mention of it in the 1760s until as late as 1787, making the comment at B167 all the more perplexing. Further, I argue that in the immediately ensuing years leading up to the publication of the *Critique of Judgment* in 1790, and in particular in relation to Johann Friedrich Blumenbach, Kant remained more ambivalent than has frequently been contended. It is not altogether clear that Kant and Blumenbach really understood the full implications of their respective positions and consequently may well have overestimated the convergence of their views.

One can only grasp in what an enormous configuration of history of science and philosophy this epigenesis model is situated, when one has reconstructed the main arguments of the fiercely conducted debate over the phenomenon of generation after the middle of the eighteenth century. What emerges is that answers to this question of how one should think about the biological origin of humans are not just some scientific paradigms among many but without question comprehend, in their speculative disclosure of abysses *[Abgründlichkeit]*, the discourses of theoretical and even practical philosophy. (Helmut Müller-Sievers, *Epigenesis: Naturphilosophie im Sprachdenken Wilhelm von Humboldts* [Paderborn: Schöningh, 1993], 29.)

Introduction

"The system of epigenesis does not explain the origin of the human body, but says far more that we don't know a thing about it."[1,2] These words of Immanuel Kant represent what I take to be the essential stance he took on epigenesis across his philosophical career, and they problematize profoundly not only his famous analogy at B167 of the *Critique of Pure Reason* (1787), but also the widely celebrated formulation in his *Critique of Judgment* (1790).[3] In a word, Kant was never fully comfortable with the idea of epigenesis. Of course, there is remarkably little consensus about exactly what epigenesis signified in eighteenth-century discourse generally, not just in Kant.[4] Accordingly, in this chapter I first describe the paradigmatic importance of epigenesis for the science of the late eighteenth century. Then I describe Kant's very problematic relationship to that theory.

1. The Eighteenth-Century Theory of Epigenesis

Epigenesis was a theory of generation giving expression to the fundamental eighteenth-century intuition of hylozoism. All the impasses of discourse in the linked spheres of metaphysics, physical theory, biology, and anthropology came to be bound up in the problem of hylozoism. That is: what properties could intelligibly be ascribed to matter, and how would this explain such questions as the causal relations of distinct substances, the principles of action at a distance, chemical attraction, electricity and heat, the mysteries of biological generation, and the mind-body relation? Eighteenth-century natural science pursued the most speculative hypotheses Newton felt prepared to interject into later editions of his *Opticks*, centered around the properties that could legitimately be considered inherent in particulate matter.[5] Scientists dwelled on attraction and repulsion, on chemical and electrical phenomena, and as they did so, they began to redefine the properties of the physical world in such a way that the notion of inert matter, and with it the impact model of force, came to seem entirely inadequate. Scientific inquiry shifted away from mathematical kinematics to what was called "experimental physics" and "natural history"—respectively, the problems of "imponderable fluids," such as electricity, magnetism, chemical bonding, light, and heat, on the one hand, and the problems of "organized form" or life, on the other.[6]

This expanded physical language had clear and crucial metaphysical concomitants. Above all, it imputed to nature a vastly grander dynamism, spontaneity, and mutability. There were three decisive frontiers of inquiry, three breaking points in the continuum of general scientific theory. First, there was the divide between the organic and the inorganic, "life" itself. Second, there was the distinction between animals and man, the question of "spirit" or "reason" (or *language*). Finally, there was the internal problem in man himself, the relation of the mind to the body, the question of "soul." From Descartes forward these boundary problems had become acute. By the mid-eighteenth century the issues surrounding "animal soul" had taken command of intellectual discourse.[7] Somehow life, spirit, and soul had to be reinterpreted, however fallibly and contingently, for "experimental science." The inspiration was offered by John Locke, though inadvertently, only as a token of the finitude of human knowledge. He made the simple point that since we could know nothing of real essences, we were not entitled to debar the possibility that God could endow matter with the power of thought.[8] John Yolton has documented how Locke's conjecture about "thinking matter" ran like a red thread through eighteenth-century philosophical discourse in both Britain and France.[9] Karl Figlio makes the key point succinctly: "The nature of the soul was in principle unknowable, but so was the nature of any substance or force."[10] "Nominal essences" would have to do since "real essences," as Locke argued, were not accessible to human understanding.[11] If one made the transition to this "nominal" register, however, if one were content to settle for what "observation and experience" could document, one could, to be sure, never claim *absolute* truth, but one could propose intersubjectively confirmable generalizations, albeit contingent and fallible.[12] The crucial category invoked along these lines was "vital materialism," the idea of emergent order as an inherent potentiality in nature itself.[13] Through it, the boldest minds of the eighteenth century proposed to explain the continuity between the living and the inert by rendering "force" immanent in the physical world. Through it, as well, they proposed to explain the continuity of animal and man using comparative anatomy to access comparative physiology.[14] Finally, through it, they proposed to explain the continuity from body to mind through the analysis of nervous response and "material ideas."[15]

The overthrow of preformation by epigenesis in life science marked the paradigm shift to vital materialism. The debate between preformation and epigenesis in the eighteenth century is well known to have occasioned both metaphysical and methodological controversies over the relation of mechanism to animism.[16] The idea of preformation had appealed to the prior age primarily because it "was consistent with the period's religious beliefs" and "solved a number of philosophical difficulties."[17] In effect, preformation removed the whole of organic life from the sphere of nature and transferred it to the original divine act of creation. Generation ceased to be a scientific problem. Preformation signified avoidance of science, a stipulative denial of the very possibility of a *life* science. But preformation nevertheless faced serious empirical anomalies. Not only could it not comfortably explain inheritance of traits from both parents, or monstrous births, or hybrid species, but it also implied empirical verification of its postulated animalcula under the increasingly effective gaze of embryological microscopy.

Epigenesis, by contrast, was the effort in life science to discern, to describe, and at least empirically to account for the immanent capacity ("force") of nature to transform itself, to construct higher plateaus of order, spontaneously. Modern usage began with William Harvey's 1651 text, *On Generation*, in which he characterized as epigenesis the characteristic of an organism that "all its parts are not fashioned simultaneously, but emerge in their due succession and order. . . . For the formative faculty . . . acquires and prepares its own material for itself."[18] First, Harvey's concept stressed *sequential emergence*, and second, it stressed *self-organization*. Spontaneity and systematicity were thus central features. Crucially, Harvey and his early eighteenth-century successors, Maupertuis and Buffon, believed that epigenesis could be assimilated to a materialist approach to science and that it utilized mechanisms, even if it could not be reduced to mechanism. Buffon and Maupertuis launched the eighteenth-century revolt against preformation by reasserting epigenesis.[19] Buffon postulated that environmental factors could trigger the epigenetic principles he characterized in terms of *moules intérieurs* and *molécules organiques*, allowing emergent change "within the reproductive process itself."[20] Buffon's *moule intérieur* was a reformulation of Harvey's formative faculty, an emergent principle of design that set in motion determinate mechanisms of organic development.[21] Buffon invoked an analogy between his "microforce" and Newton's gravity to ease the epistemological problem of accessing "real" causes.[22] That became a consistent practice among all subsequent theorists of epigenesis.

Ironically enough, the path-breaking work on irritability and sensibility of the leading preformationist, Albrecht von Haller, contributed substantially to the very methodology that he found unacceptable when called on in support of epigenesis.[23] Caspar Friedrich Wolff made that case in what is taken to be the most important reformulation of epigenesis in the mid-eighteenth century. His *vis essentialis* was conceived as a Newtonian force that induced through certain chemical processes the production of organic matter out of inorganic matter in accordance with regular and empirically demonstrable patterns.[24] Johann Gottfried Herder was drawing directly on Wolff's work in articulating his idea of epigenesis in 1784, though he may have been aware as well of Johann Friedrich Blumenbach's work on the *Bildungstrieb* (1781), which marked the ultimate triumph of epigenesis in eighteenth-century science.[25]

Arthur Genova identifies three crucial elements in the concept of epigenesis in its full-fledged form in the eighteenth century: *autonomy*, *community*, and *reflexivity*.[26] In my terms, I would stress the radicality of emergence by replacing autonomy with spontaneity. By *community* Genova signifies the mutuality of cause and effect and of parts and whole that is central to the notion of organic form, especially as Kant articulated it. *Reflexivity*, finally, has to do with the self-regulating, self-forming dimension as a persistent feature of life-forms, over and above the question of their emergence de novo. Each of these elements poses decisive challenges, methodologically and metaphysically, to a physical science on the sort of Newtonian foundations Kant preferred. At the metaphysical end of this spectrum lie the problems of *radical novelty* in a model that stresses systematic causal determination and, conversely, of the persistent *determinacy* of specific life forms: why there is so much regularity in a context of apparently radical freedom.[27] At the methodological end of the spectrum lie the questions of empirical verification and of the degree and nature of mechanical execution of the self-organizing principle in the life forms.

Kant's commitments impeded his recognition of these recent developments in eighteenth-century science, distancing him from some of its most creative and effective currents.[28] His refusal to consider these possibilities must be associated with his views not merely about *method* but especially about *metaphysics*. Kant had metaphysical positions to defend: the traditional notion of a transcendent, intelligent Deity who created the world ex nihilo, and the notion of individual moral responsibility, which in his view required man to have at least *noumenal* freedom. As he saw it, the "materialist" and "pantheist" trends in science and cosmology, above all the renaissance of the philosophy of Spinoza in Germany, threatened these positions.

There were few ideas Kant struggled to keep divided more than life and matter. Kant denied that we could think of nature as alive: "the possibility of living matter cannot even be thought; its concept involves a contradiction, because lifelessness, *inertia*, constitutes the essential character of matter."[29] He elaborated: "life means the capacity of a substance to determine itself to act from an internal principle, of a finite substance to determine itself to change, and of a material substance to determine itself to motion or rest as change of its state."[30] Kant defended an idea that Descartes first proposed for physics and that Newton ostensibly maintained in his great works, namely, *inert* matter.[31] Consequently, there were two boundaries Kant sought ceaselessly to enforce. First, he wished to secure the distinction of organic life from the inorganic, affirming the uniqueness and mystery of organisms as phenomena of empirical nature, and upholding the utter inexplicability of the origins of life.[32] He repeatedly claimed that there could never be a Newton of the blade of grass.[33]

Second, Kant insisted on a distinction of man from the rest of organic life. The only power capable of self-determination, Kant emphasized, was intelligent will. Intelligent will could never be found in phenomena; it was not part of nature. It was a noumenal property. Even man, in part a being of the natural order, had "life" only by virtue of his other, noumenal aspect. Kant's insistence on the uniqueness of reason and freedom necessitated the categorical separation of man from other animals. This categorical separation motivated Kant to insist on the irrevocable fixity of all species (and even of human races).

In a series of writings from 1784 to 1790 Kant struggled to secure natural science against what he took to be wildly speculative notions proliferating in the emergent sciences that we now call chemistry and biology. *Metaphysical Foundations of Natural Science*, written in the summer of 1785, articulated these concerns: first, it asserted categorically the distinction of life from matter; second, it developed a restrictive formulation of Newtonian force; but, above all, it prescribed limits for scientific investigation according to what Kant viewed as sound Newtonian method. The upshot was Kant's striking pronouncement, in the Preface to *Metaphysical Foundations of Natural Science*, that the new fields of "experimental physics"—chemistry and the life sciences—were "improper sciences" (*uneigentliche Wissenschaften*).[34]

Kant adamantly denied the notion of nature as active "in-and-for-itself":

> We perhaps approach nearer to this inscrutable property if we describe it as an analogon of life, but then we must either endow matter, as mere matter, with a property which contradicts its very being (hylozoism) or associate therewith an alien principle standing in communion with it (a soul). But in the latter case we must, if such a product is to be a natural product, either presuppose organized matter as the instrument of that soul, which does not make the soul a whit more comprehensible, or regard the soul as artificer of this structure, and so remove the product from (corporeal) nature.[35]

The philosophical problem, Kant insisted, allowed only one solution: a transcendent creator. In Kant's words, "Nature is no longer estimated as it appears like art, but rather insofar as it actually *is* art, though superhuman art."[36] This conjecture of a Nature-for-God came to formulation via the analogy of purposiveness. "Our reason has in its power for the judgment no other principle of the possibility of the object, which it inevitably judges teleologically, than that of subordinating the mechanism of nature to the architectonic of an intelligent Author of the world."[37] Such "physico-theology" in the form of a "Technic of Nature" was inevitable for man's discursive understanding, Kant claimed. Of course, he formulated all this as a *heuristic for inquiry*, not an *ontology of nature*: that is the critical "purity" preserving Kant from "dogmatism."

2. Kant's Persistent Ambivalence toward Epigenesis

From the beginning, Kant was acutely sensitive to the whole constellation of concerns, methodological and metaphysical, surrounding epigenesis. Already in his *Only Possible Argument for a Demonstration of the Existence of*

God (1763), Kant showed his awareness of the new twist toward epigenesis introduced by Maupertuis and Buffon.[38] When Kant turned to questions of life science in his first essay on race (1775/77), he adopted eagerly Albrecht von Haller's modified preformation theory both because it seemed more methodologically viable and also—perhaps even more—because it reasserted with full rigor the metaphysical objection against hylozoism.[39] Bonnet and Haller developed the sophisticated version of preformation in the early 1760s in response to the challenge first of Maupertuis and Buffon and then, more fundamentally, of Caspar Friedrich Wolff.[40] As Günter Zöller characterizes this form, "preformationism is primarily a theory concerning the generation of distinct parts (organs) in the growing embryo. It maintains that growth is quantitative growth of preexisting parts . . . no qualitative embryological growth or formation of new parts." In that light the term *Anlagen* had a quite specific application, just as Kant articulated it in his first essay on race, namely, to "'conditions of a certain development . . . insofar as the latter only concerns the size and the relation of parts' . . . [as] opposed to germs ['*Keime*'], which are conditions for the development of new parts."[41] That is, the role of *Anlagen* could be construed in a quasi-mechanistic fashion; the essential metaphysical principle guaranteeing species difference (and persistence) was assigned to *Keime*. Kant also believed that he could advance the argument, both in the formulation of the mechanism of adaptation and variation—the great weakness of earlier preformation theories—and also in his general methodological idea of "natural history."[42]

Thus, by the time he published the first *Critique* in 1781, Kant considered himself sufficiently adept in the theory of generation to offer a telling analogy to his theory of knowledge:

> I understand under the "Analytic of Concepts" . . . the still little investigated dissection of the capacity of the understanding itself, in order thereby that we search into the possibility of a priori concepts, seeking them out in the understanding alone, as their source of birth. We will therefore follow the pure concepts up to their first germs and capacities [*Keimen und Anlagen*] in the human understanding, in which they lie predisposed, until they finally, on the occasion of experience, develop and through exactly the same understanding are displayed in their purity, freed from their attending empirical conditions.[43]

This analogy of 1781, as Phillip Sloan has established, is crucial to any assessment of the more famous analogy of 1787 to epigenesis.[44] First, the 1781 language is unequivocally a *preformationist* analogy. The concepts lie "predisposed" in the understanding; they are not produced, they are occasioned. Moreover, Kant meant to suggest an element in the analogy would be central to his thinking throughout, namely, that just as *Keime* and *Anlagen* were inaccessible to ultimate derivation, so too the concepts of the understanding were simply givens behind which we could not seek. The clearest formulation is in the revised version (1787) of the first *Critique*:

> This peculiarity of our understanding, that it can produce a priori unity of apperception solely by means of the categories, and only such and so many, is as little capable of further explanation as why we have just these and no other functions of judgment, or why space and time are the only forms of our possible intuition.[45]

This view was fully entailed in the earlier version, with its speculative gesture to a common but unknown root of sensibility and understanding.[46]

The critical point is that Kant wished to see an analogy between *preformation* and transcendental philosophy, and not between the latter and *epigenesis*. Indeed, there was no affirmation of epigenesis in any of Kant's writings before 1787.[47] In his dispute with Herder and Forster from 1784 through 1788, that is, up through the time of his revision of the first *Critique*, Kant remained committed to preformation. Indeed, the insistence on limiting adaptive change in organisms was the decisive point in Kant's critique of Herder's generation theory:

> As the reviewer understands it, the sense in which the author uses this expression [i.e., *genetische Kraft*] is as follows. He wishes to reject the system of evolution on the one hand, but also the purely mechanical influence of external causes on the other, as worthless explanations. He assumes that the cause of such differences is the vital principle [*Lebensprinzip*] which modifies itself from within in accordance with variations in external circumstances, and in a manner appropriate to these. The reviewer is fully in agreement with him here, but with this reservation: if the cause which organizes from within were limited by its nature to only a certain number and degree of differences in the development of the creature which it organizes (so that, once these differences were exhausted, it would no longer be free to work from another archetype [*Typus*] under altered circumstances), one could well describe this natural development of formative nature in terms of germs [*Keime*] or original dispositions [*Anlagen*], without thereby regarding the differences in question as originally implanted and only occasionally activated mechanisms or buds [*Knospen*] (as in the system of evolution); on the contrary, such differences should be regarded simply as limitations imposed on a self-determining power, limitations which are inexplicable as the power itself is incapable of being explained or rendered comprehensible.[48]

Clearly, Kant was invoking *Keime* in the sense of Haller's sophisticated preformationism against what he saw as an insupportable hylozoism in Herder.

In his second essay on race in 1785, Kant reiterated his fundamental principles, making even more explicit his commitment to the fixity of species.

"Throughout organic nature, amid all changes of individual creatures, the species maintain themselve unaltered [*die Species derselben sich unverändert erhalten*]." Kant went on to argue that this was an essential principle of scientific investigation, without which every concept would dissolve:

> [I]f some magical power of the imagination . . . were capable of modifying . . . the reproductive faculty itself, of transforming Nature's original model or of making additions to it, . . . we should no longer know from what original Nature had begun, nor how far the alteration of that original may proceed, nor . . . into what grotesqueries of form species might eventually be transmogrified [*in welche Fratzengestalt die Gattungen und Arten zuletzt noch verwildern dürften*]. . . . I for my part adopt it as a fundamental principle to recognize no power . . . to meddle with the reproductive work of Nature . . . [to] effect changes in the ancient original of a species in any such way as to implant those changes in the reproductive process and make them hereditary.[49]

When we read Kant's highly charged language we cannot but discern that again it is the idea of *hylozoism*—of any radical spontaneity in matter itself—that Kant could not abide. All organic form had to be fundamentally distinguished from mere matter. "Organization" demanded separate creation. Eternal "inscrutability" was preferable to any "speculative" science. In the third *Critique* Kant would twice insist that no human could ever achieve a *mechanist* (he meant, as well, an immanent, physicalist, or *materialist*) account of so much as a "blade of grass."[50] Organisms, as empirically given forms of nature, became literally indecipherable once the concept *life* was removed, leaving us to grope after them by analogies. Kant remained adamant that the *ultimate* origin of "organization" or of formative force required a *transcendent-metaphysical*, not a physical, account. "Chance or general mechanical laws can never bring about such adapation. Therefore we must see such developments which appear accidental according to them, as *predetermined* [*vorgebildet*]." External factors could be occasions, but not direct causes of changes that could be inherited through generation. "As little as chance or physical-mechanical causes can generate [*hervorbringen*] an organic body, so little will they be able to effect in them a modification of their reproductive powers which can be inherited."[51] Above all, "how this stock [of *Keime*] arose, is an assignment which lies entirely beyond the borders of humanly possible *natural philosophy*, within which I believe I must contain myself," Kant wrote in 1788.[52]

What drew Kant to epigenesis at all? The first answer is that Kant appropriated the term from Herder.[53] He found it in a form that was too radical for his taste, yet he believed that he could seize it from Herder and make it stand precisely for his own position. All that required was a two-step process. First, Kant had to insist that even epigenesis implied preformation: at the origin there had to be some "inscrutable" (transcendent) endowment, and with it, in his view, some determinate restriction in species variation. Thereafter, the organized principles within the natural world could proceed on adaptive (mechanical) lines. This made *epigenesis* over into Kant's variant of *preformation*. Even so, this seemed to postulate the objective *actuality* of these forces for natural science. That violated Kant's "Newtonianism." Hence Kant faced the ultimate need for a second step: to transpose the whole matter from the constitutive to the regulative order.

In his *Metaphysics Lectures* Kant left us some crucial evidence regarding how he conceived the juxtaposition of *preformation* and *epigenesis*.[54] The student notes are not entirely coherent, but we can make certain clear inferences from the passages. First, Kant found the contrast of *educt* and *product* crucial for his conceptualization.[55] The difference between them is that in an educt all the relevant material preexists, and only its aggregation is shuffled, whereas in a product, altogether new things emerge, presumably by immanent processes (*per traducem*). Kant saw this as a mode of thought already established in chemistry and he clearly saw the theories of generation in the life sciences as variants of the same method of conceptualization. Thus there were, for him, only *two* theoretical possibilities for the generation of bodies (or souls), namely, preformation (the educt-theory) and epigenesis (the product-theory). Kant presented epigenesis in both sets of notes as a *hypothetical*, not an *assertoric* judgment: *if* we have grounds for assuming the epigenesis theory, *then* . . . Clearly, Kant was not committing himself to the hypothesis; he was not saying "*since* we have grounds." Indeed, if we are attentive to both passages, what emerges is that Kant in fact *rejected* this hypothesis, and therefore rejected epigenesis, as he made clear especially at the close of the second passage: "there is no alternative but to assume the soul is preformed."[56] My point is that in all this material there is still nothing like an unequivocal affirmation of epigenesis.

How are we to understand what Kant had in mind by claiming his critical philosophy was "a system, as it were, of the *epigenesis* of pure reason," the famous passage at B167 in the 1787 version of the first *Critique?* That Kant found it appropriate to draw an analogy of his own transcendental method in philosophy to epigenesis in embryology suggests that something very central was involved for him in this issue in the life sciences.[57] Indeed, spontaneity and systematicity, two crucial ideas in Kant's theory of reason, find empirical analogs in the idea of epigenesis in nature. But we must be sensitive to the uses of analogy that Kant was prepared to acknowledge, as Hans Ingensiep has argued. Ingensiep suggests that Kant did not intend by analogy to extend his formal argument for transcendental philosophy, nor was analogy serving here as a heuristic to enable further discoveries (as in the Kuhnian sense of paradigm); rather, it was only for *"intuitive illustration"* [*anschaulichen Verdeutlichung*].[58] At most, Kant gestured to "structural similarities," and "accordingly in no way can it be construed as a claim for any compelling ontological connection between the respective philosophical and biological positions."[59] But this does not yet clarify the tension between the concepts of epigenesis and preformation as they featured in Kant's thinking.

How exactly did Kant employ this analogy of transcendental philosophy to *epigenesis* at B167 of the first *Critique*? The argument of §27 of the Transcendental Deduction in B (which includes the passage at B167) is an elaboration of the argument of §36 of the *Prolegomena* (1783).[60] Both arguments present a disjunctive judgment: either experience generates the categories or the categories generate experience. Since Kant stipulated that we *already* knew that the categories had to be a priori, only the second option was really available. In the B Deduction, he drew a new analogy to *generatio aequivoca*—spontaneous generation—already an exploded idea in the natural science of the day.[61] Note that the fundamental analogy structure at B167 invokes the disjunction: *either* spontaneous generation *or* epigenesis; preformation is introduced in connection with the misguided endeavor to insert a third, intermediate position.[62] What I want to highlight first is Kant's use of preformation with a clearly *negative* connotation. Kant's whole point against the intermediate position was that we need a stronger bond between the categories and experience if we are to take seriously the necessity that is the essence of transcendental grounding. That bond could be achieved only if it were *self-formed*, not endowed, even by God. That is why he italicized the strange term *self-thought* [*selbstgedacht*], in his phrase "*self-thought* first principles a priori."[63] That is why Kant suddenly invoked the idea of *epigenesis*. But that still does not resolve the problem.

What could Kant possibly have been thinking at B167? Why, for the first time, would he have put preformation in a negative context and epigenesis in a remarkably and unprecedentedly positive one? In terms of the educt/product distinction, we have a clearer sense of what Kant thought the essential point of epigenesis was. But we also see it as problematically creative or spontaneous from Kant's vantage, as ascribing too much power to mere created substances. That is, the metaphysical issue with epigenesis was still hylozoism. Was there something that Kant now saw in the idea of epigenesis that could help him to elucidate the peculiar and essential spontaneity of the understanding in his transcendental deduction? Kant wanted to stress the difference between a Leibnizian sense of the innate *capacities* of mind and a Cartesian sense of innate *ideas*.[64] The categories themselves should be seen not as preformed, but only as produced spontaneously by an innate *capacity* or *power*—a "faculty" of mind, whose own origin was utterly "inscrutable." Spontaneity of the categories was not sufficient for Kant's transcendental deduction; he also needed their constitutive sovereignty over experience. The ordering force of the innate ("epigenetic") powers of mind had to be efficacious in empirical experience; it had to be able to *produce* new knowledge ("synthetic a priori judgments"). That is, it had to be a *real* cause (of knowledge), though a cause in a sense different from what would be asserted within specific empirical judgments regarding sensible intuition. Kant's epigenesis analogy, in short, built intellectual causation (determination, constitution) into the fundamental structure of the transcendental deduction of the possibility of experience.

If epigenesis needs to be understood on the model of a product, what were the necessary preconditions for immanent emergence? We have reason to suspect that Kant—however clear he may have been about what he wanted to accomplish in the transcendental deduction—may not have grasped clearly what he was playing with in the analogy to epigenesis. Stealing it from Herder may have gratified him; it may even have led him to an increased clarity about the sort of spontaneity he needed for the origination and systematicity of his categories. But how was he to square his reservations with the revolution in thinking about epigenesis inaugurated by Blumenbach in 1781? We need to consider Kant's response to Forster, "On the Use of Teleological Principles" (1788), in this light. Kant referred to Blumenbach in a footnote, invoking the first edition of the *Handbuch der Naturgeschichte* (1779), which Kant owned.[65] As we know, Blumenbach revolutionized his thought shortly after publishing that work, developing his theory of the *Bildungstrieb* in 1780/81. In that new work, Blumenbach strongly repudiated *any sense* of germs [*Keime*].[66] In Kant's same note, interestingly, we find added the observation: "this insightful man also ascribes the *Bildungstrieb*, through which he has shed so much light on the doctrine of generation, not to inorganic matter but solely to the members of organic being."[67] Thus, Kant had *some* acquaintance with Blumenbach's term already in the fall of 1787 when he composed the essay.[68]

The important point, however, is that Kant did not alter his own theory in any significant measure in his essay of 1788. Most tellingly, he persisted in his use of *Keime*. When we ask after the specific point for which Kant actually invoked Blumenbach, it was to dismiss what in the *Critique of Judgment* he would call a "daring adventure of reason," namely, the transformation of the great chain of being from a taxonomy to a phylogeny, which had been raised by Forster.[69] Kant was happy to report that this "widely cherished notion preeminently advanced by Bonnet" came under appropriate criticism in Blumenbach's *Handbuch der Naturgesichte*.[70] Indeed, Blumenbach shared Kant's skepticism about the genetic continuity of life forms. What bound them most together was their commitment to the fixity of species. But how could the transmutationist implications of epigenesis be contained within the limits of the fixity of species? This was the essential question that Kant had posed in his second essay on race in 1785, and the stakes of the question were not small. Without some regulation in the history of generation, the prospect of the scientific reconstruction of the connection between current and originating species (*Naturgeschichte*, in Kant's new sense, or the "archaeology of nature," as he would call it in the third *Critique*) would be altogether hopeless. But it was not simply a *methodological* issue, however dire. There was also an essential *metaphysical* component: *hylozoism*.

In 1789 Blumenbach sent Kant a copy of the second edition of his essay on the *Bildungstrieb*, one that not only expounded its epigenetic aspects but also set it in methodological terms that showed clearly the influence of Kant's own arguments about distinguishing mechanical from teleological explanation. Blumenbach advised readers to disregard his earlier, "immature" formulations.[71] But which differences did he really introduce? Perhaps most prominent was an explicit Newtonian analogy.[72] Second, as noted, Blumenbach showed awareness of the teleology/mechanism problem that Kant highlighted in the 1788 essay. That is, in the 1789 version, Blumenbach was self-consciously assimilating his methodological presuppositions to Kant's. Above all, Blumenbach repudiated hylozoism: "No one could be more totally convinced by something than I am of the mighty abyss which nature has fixed [*befestigt*] between the living and the lifeless creation, between the organized and the unorganized creatures."[73] This was what Kant found most gratifying in the new book, as he reported in his letter of acknowledgment to Blumenbach.[74] In the *Critique of Judgment* he elaborated:

> Blumenbach . . . rightly declares it to be contrary to reason that raw matter should originally have formed itself in accordance with mechanical laws, that life should have arisen from the nature of the lifeless, and that matter should have been able to assemble itself into the form of a self-preserving purposiveness by itself; at the same time, however, he leaves natural mechanism an indeterminable but at the same time also unmistakable role under this inscrutable principle of an original organization, on account of which he calls the faculty in the matter in an organized body (in distinction from the merely mechanical formative power [*Bildungskraft*] that is present in all matter) a formative drive [*Bildungstrieb*] (standing, as it were, under the guidance and direction of that former principle).[75]

The leading life scientist of the day seemed to be affirming just the same metaphysical and methodological discriminations that Kant himself demanded.

By the time Kant came to write the crucial passage in the *Critique of Judgment*, then, we can presume that he was aware of Blumenbach's sophisticated theory of epigenesis. One indication, as Phillip Sloan has noted, was that Kant suppressed any mention of *Keime* in that work, though it still thronged with the term *Anlage*.[76] But what progress had Kant made on the conundrum of preformation versus epigenesis? It is important to distinguish two quite distinct sets of discriminations in the *Critique of Judgment* that both point back to B167 but have different implications. The first discriminations come in a footnote to §80; the second come in the main text of §81. The footnote to §80 evokes the familiar term *generatio aequivoca* in order, as before, to disparage it. The contrast, however, is not to epigenesis or to preformation, but rather to *generatio univoca*, which Kant further subdivides into *generatio homonyma* and *generatio heteronyma*. While spontaneous generation was once again dismissed as contradictory, Kant asserted that transmutation of species (*generatio heteronyma*) was not contradictory, only unfound in experience. Thus, the issue at stake in this discrimination is the principle of the persistence of species. In §81, however, we come upon a different schematization. Here, Kant postulated that we must think of organisms on the analogy of an intelligent creation and that when we do so we face alternatives that can best be grasped *in terms drawn from metaphysics* (i.e., the obverse of the analogy at B167). The categories Kant offered were *occasionalism* and *prestabilism*. He dismissed occasionalism as curtly as he had dismissed spontaneous generation (though, of course, for different reasons), and in turning to "prestabilism" he distinguished two subsets: *individual preformation*, which he identified with the "theory of evolution" (i.e., encapsulation) and termed an "educt," and *generic preformation*, which Kant suggested was the proper sense of *epigenesis*. That is, while a "product," epigenesis "still performed in accordance with the internally purposive predispositions that were imparted to its stock."[77] Hence even as he was prepared to advocate *epigenesis*, Kant set strict limits on it: ultimately this was still just "generic preformation," that is, it, too, required the intervention of a *transcendent* causality.[78] What attracted him to epigenesis over individual preformation, Kant averred succinctly, was that it entailed "the least possible application of the supernatural" in scientific theory.[79]

This is the decisive passage and it requires careful exegesis. First, it is apparent that Kant reconfigured his whole conceptualization *under the aegis of preformation*. Second, there is no strict parallelism between the distinctions of §80 and §81: the distinction between *generatio homonyma* and *generatio heteronyma* does not map neatly onto that between individual and generic preformation. That suggests that a different point is being made in the latter distinction, and indeed this point has to do with the character of the *causality* that must be employed in conceptualizing organic forms altogether, namely, the inadequacy not merely of *mechanism* but above all of *materialism*.[80] Yet there is at least a measure of spillage between the two patterns of discrimination, for Kant found the idea of the transmutation of species—*generatio heteronyma*—to induce the very sorts of loose thinking in science that might read *epigenesis* as hylozoism, as a *vital* materialism. It was this above all that he wished to circumvent, both with his ontological argument that even epigenesis depended on an original creation that instilled organization into inert matter and with his methodological argument that an empirical science of life forms could work only with maxims of reflective judgment imputing purposiveness, and thus that the very idea of a natural purpose was merely a heuristic fiction suited to our limited reason.

But what is also clear, as Sloan has argued, is that Kant had still not come to terms with the implications for his analogy between epigenesis and transcendental philosophy.[81] If epigenesis signified what Blumenbach was urging, then the security of *Keime*, with their determinate restrictions on species change in biology, would have to be forsaken. And, by analogy, the implications for Kant's transcendental grounding of the categories would be similarly grave. Of course, Kant's escape was to suggest an epistemological evasion of this unpalatable ontological prospect. Zumbach phrases it suitably: "Kant's claim that there are free causes in living processes is elliptical. He is actually claiming that living processes must be viewed in terms of the *idea* of a free cause."[82] That is an *epistemological* strategy, a heuristic, not a fact. In Kantian terms, there is a subjective necessity—a "need of reason"—or this move, but no objective necessity, no natural law evident in the matter at hand (the "order of nature").[83] This is the famous argument of Kant's Dialectic of Teleological Judgment, and his resolution is that in order to make organic forms intelligible at all, we must have recourse to the analogy of purpose or design.[84] Kant transposes his metaphysical problem into a methodological one, his ontological need into an epistemological constraint: "nature [i.e, the "order of nature" as a system] can only be understood as meaningful if we take it at large to be designed."[85] That is, "we need to be able to comprehend all of nature, not as a living being, but as a rational analog of a living being."[86]

Epigenesis incites a fundamental erosion of Kant's boundary between the constitutive and the regulative, between the transcendental and the empirical: a naturalism beyond anything Kant could countenance, though his own thought carried him there. With epigenesis, the "order of nature" is greater than the order of Kant's version of Newtonian physics, and the paradigm for science necessarily exceeds the "Newtonian" constraints Kant wished to impose on it.[87] His demand that the life sciences submit to the methodological principles of his "Newtonianism" (as in his critique of Herder and his dispute with Forster, and above all in his Preface to the *Metaphysical Foundations of Natural Science*) was misguided.[88]

To be consistent, what Kant had to do was to *qualify* his conception of "Newtonian" science in order to make room for the ontological possibility of life forces.[89] According to Arthur Lovejoy, Kant

> recoils in horror before the idea of admitting that real species are capable of transformation ... because of certain temperamental peculiarities of his mind—a mind with a deep scholastic strain ... one that could not quite endure the notion of a nature all fluent and promiscuous and confused, in which series

of organisms are to an indefinite degree capable of losing one set of characters and assuming another set. He craved, above all, a universe sharply categorized and classified and tied up in orderly parcels.[90]

Lovejoy is often intemperate in his criticism of Kant, but there is more than a germ of truth in this passage.[91] What Kant refused to warrant was the overweening intuition of the epoch, that, in Frederick Gregory's formulation, "nature was not a timeless and immutable machine, but a temporal and developing organism."[92] Goethe gave expression to this when he tried to explain how he reacted to Linnaeus: "what he wanted to hold apart by force I had, according to the innermost need of my nature, to strive to bring together."[93]

Notes

1. Kant, *Vorlesungen über Metaphysik*, Ak. 29, 761.
2. This is a substantially abridged version of my essay, "'This inscrutable *principle of an original organization*': Epigenesis and 'Looseness of Fit' in Kant's Philosophy of Science," *Studies in History and Philosophy of Science* 34 (2003): 73–109.
3. All translations of Kant are by the author.
4. Shirley Roe, "Rationalism and Embryology: Caspar Friedrich Wolff's Theory of Epigenesis," *Journal of the History of Biology* 12 (1979): 1–43, citing 3n; Helmut Müller-Sievers, *Epigenesis: Naturphilosophie im Sprachdenken Wilhelm von Humboldts (Humboldt-Studien)* (Schöningh, 1993); Müller-Sievers, *Self-Generation: Biology, Philosophy and Literature around 1800* (Stanford: Stanford University Press, 1997); Zammito, "Epigenesis: Concept and Metaphor in Herder's *Ideen*," in *Vom Selbstdenken: Aufklärung und Aufklärungskritik in Johann Gottfried Herders "Ideen zur Philosophie der Geschichte der Menschheit,"* ed. Rudolf Otto and John Zammito (Heidelberg: Synchron, 2001), 131–45.
5. P. M. Heimann and J. E. McGuire, "Newtonian Forces and Lockean Powers: Concepts of Matter in Eighteenth-Century Thought," *Historical Studies in the Physical Sciences* 3 (1971): 233–306; P. M. Heimann, "Voluntarism and Immanence: Conceptions of Nature in Eighteenth-Century Thought," *Journal of the History of Ideas* 39 (1978): 271–83; P. M. Heimann, "'Nature is a perpetual worker': Newton's Aether and Eighteenth-Century Natural Philosophy," *Ambix* 20 (1973): 1–25; Peter Harman, *Metaphysics and Natural Philosophy: The Problem of Substance in Classical Physics* (Brighton, Sussex: Harvester Press; Totowa, NJ: Barnes & Noble Books, 1982); Arnold Thackray, *Atoms and Powers: An Essay on Newtonian Matter-Theory and the Development of Chemistry* (Cambridge, MA: Harvard University Press, 1970); John Yolton, *Thinking Matter: Materialism in Eighteenth-Century Britain* (Minneapolis: University of Minnesota Press, 1983); Margaret Jacob, *The Radical Enlightenment: Pantheists, Freemasons and Republicans* (London: Allen and Unwin, 1981), esp. 1–64.
6. Michael Friedman notes this in *Kant and the Exact Sciences* (Cambridge, MA: Harvard University Press, 1992). For a rich characterization of this emergent science, see the essays of Peter Hans Reill, "Science and the Science of History in the

Spätaufklärung," in *Aufklärung und Geschichte*, ed. Hermann E. Boedeker et al. (Göttingen: Vandenhoeck and Ruprecht, 1986); "Anti-Mechanism, Vitalism and Their Political Implications in Late Enlightened Scientific Thought," *Francia* 16, no. 2 (1989): 195–212; "Science and the Construction of the Cultural Sciences in Late Enlightenment Germany: The Case of Wilhelm von Humboldt," *History and Theory* 33, no. 3 (1994): 345–66; and "Anthropology, Nature and History in the Late Enlightenment: The Case of Friedrich Schiller," in *Schiller als Historiker*, ed. Otto Dann et al. (Stuttgart: Metzler, 1995), 243–65. See also my chapter "Kant versus Eighteenth Century Hylozoism," in *The Genesis of Kant's Critique of Judgment* (Chicago: University of Chicago Press, 1992), 189–99, and the literature cited therein.

7.   "The concept of the animal soul did not give rise to any serious problems until the seventeenth century, when Cartesian dualism brought out distinctions which had been latent in the dominant Aristotelian tradition. Aristotle had postulated graduations from inert, inanimate matter to plants, which had the additional functions of nourishment and reproduction, to animals, which were also endowed with sensation, motion, and all degrees of mental functions except reason: he reserved reason for man" (Robert Young, "Animal Soul," *Encyclopedia of Philosophy*, vol. 1, pp. 122–27, citing 122).

8.   Locke, *Essay Concerning Human Understanding*, 4:3:6; see Yolton, *Thinking Matter*, 14–17.

9.   Yolton, *Thinking Matter*; Yolton, *Locke and French Materialism* (Oxford: Clarendon Press, 1991).

10.   Karl Figlio, "Theories of Perception and the Physiology of Mind in the Late Eighteenth Century," *History of Science* 12 (1975): 183. And see Robert Smith, "The Background of Physiological Psychology in Natural Philosophy," *History of Science* 11 (1973): 75–123, esp. 79: "functional concepts have enabled scientists to develop a discipline of physiological psychology in spite of tendencies towards a dualistic ontological framework."

11.   Wolfgang Proß argues convincingly that Locke's suspicion of the adequacy (by the absolutist measure) of science weighed heavily on eighteenth-century thought ("Herder und die Anthropologie seiner Zeit," in *Herder und die Anthropologie der Aufklärung*, ed. Wolfgang Proß (Darmstadt: Wissenschaftliche Buchgesellschaft, 1987), 1136). And see Margaret Osler, "John Locke and the Changing Ideal of Scientific Knowledge," *Journal of the History of Ideas* 31 (1970): 3–16.

12.   The historical novelty of this notion of empirical law can be grasped only in the context of the persistence of an absolutist notion: these issues were central to the critical Kant and played out especially in the first *Critique* and in his *Metaphysical Foundations of Natural Science*.

13.   Timothy Lenoir uses the notion "vitalist materialism" to characterize the eighteenth-century life sciences in Germany in *The Strategy of Life* (Dordrecht: Reidel, 1982), but he works very hard to render it properly "regulative" in the Kantian sense. However, the very best efforts of scientists well informed about Kant's methodology of science and seeking to follow its terms all *failed* by 1800. (See my "'Method' versus 'Manner'? Kant's Critique of Herder's *Ideen* in the Light of the Epoch of Science, 1790–1820," in *Herder-Jahrbuch/Herder Yearbook 1998*, ed. Hans Adler and Wulf Koepke (Stuttgart: Metzler, 1998), 1–26.) Kant's boundary between regulative

and constitutive was simply not viable for the culture of science of the late eighteenth century. If, by his lights, all the scientists of "vital materialism" fell into the error of the constitutive employment of what should only have been regulative ideas, by *our* lights, all of them were necessarily *gesturing to the real in theoretical models*; i.e., the only thing science *can be* is "regulative" in Kant's sense, and yet what it is being "regulative" *about* is the world. Methodologically, his distinction is impotent, and metaphysically it leaves *all*, not just *some* science in limbo. In short, I suggest that our debates about "scientific realism" need to be invoked to make better sense of the methodological situation of science in the late eighteenth century.

14.   William Bynum, "The Anatomical Method, Natural Theology, and the Functions of the Brain," *Isis* 64 (1973): 445–68.

15.   Wolfgang Riedel, "Influxus physicus und Seelenstärke," in *Anthropologie und Literatur um 1800*, ed. Jürgen Barkhoff and Eda Sagarra (Munich: Iudicium, 1992), 29.

16.   Roe, "Rationalism and Embryology"; A. E. Gaissinovich, "Le rôle du Newtonianisme dans la renaissance des idées épigénetiques en embryologie du XVIIIe siècle," in *Actes du XIe Congrès International d'Histoire des Sciences* 5 (1968): 105–10; Charles Bodemer, "Regeneration and the Decline of Preformationism in Eighteenth-Century Embryology," *Bulletin of the History of Medicine* 38 (1964): 20–31; Olaf Breidenbach, "Die Geburt des Lebendigen—Embryogenese der Formen oder Embryologie der Natur?—Anmerkungen zum Bezug von Embryologie und Organismustheorien vor 1800," *Biologisches Zentralblatt* 114 (1964): 191–99; Hans-Jorg Rheinberger, "Über Formen und Gründe der Historisierung biologischer Modelle von Ordnung und Organisation am Ausgang des 18. Jahrhunderts," in *Gesellschaftliche Bewegung und Naturprozeß*, ed. Manfred Hahn and Hans-Jörg Sandkühler (Cologne: Paul-Rugenstein, 1981), 71–81; Hans-Jorg Rheinberger, "Aspekte des Bedeutungswandels im Begriff organismischer Ähnlichkeit vom 18. zum 19. Jahrhundert," *History and Philosophy of the Life Sciences* 8 (1986): 237–50; Reinhard Mocek, "Caspar Friedrich Wolffs Epigenesis-Konzept—Ein Problem im Wandel der Zeit," *Biologisches Zentralblatt* 114 (1995): 179–90; Müller-Sievers, *Epigenesis*; Müller-Sievers, *Self-generation*; V. Dawson, "Regeneration, Parthenogenesis, and the Immutable Order of Nature," *Archives of Natural History* 18 (1991): 309–21; François Duchesneau, "Haller et les théories de Buffon et C. F. Wolff sur l'épigenèse," *History and Philosophy of the Life Sciences* 1 (1985): 65–100; Elizabeth Haigh, "Vitalism, the Soul, and Sensibility: The Physiology of Théophile Bordeu," *Journal of the History of Medicine* 31 (1976): 30–41; and Zammito, "Epigenesis: Concept and Metaphor."

17.   Roe, "Rationalism and Embryology," 2.

18.   William Harvey, *On Generation* (1651; reprint: Ann Arbor: Edwards, 1943), 366.

19.   In the words of Phillip Sloan, "Whereas preformationism had rendered the relations of organisms purely occasional, Buffon's theory, relying on the immanent continuity of the *moules intérieurs* and the *molécules organiques*, required a literal material continuity of forms in relation of true generation of like by like in historical time. . . . By reinterpreting the issue of generation in epigenetic terms, Buffon provided a means by which the contingencies of geography and climate, acting upon the *molécules*, could affect the actual reproductive lineage of the species" (Sloan,

"The Gaze of Natural History," in *Inventing Human Science*, ed. Christopher Fox et al. (Berkeley: University of California Press, 1995), 133).

20. Ibid., 135.

21. Sloan, "Buffon, German Biology, and the Historical Interpretation of Biological Species," *British Journal for the History of Science* 12 (1979): 118.

22. Lenoir, "The Göttingen School and the Development of Transcendental *Naturphilosophie* in the Romantic Era," *Studies in History of Biology* 5 (1981): 123.

23. Ibid., 135.

24. Caspar Friedrich Wolff, *Theorie von der Generation* (1764; reprint: Stuttgart: G Fischer, 1966); Roe, "Rationalism and Embryology"; Gaissinovich, "Le rôle du Newtonianisme"; Richard Aulie, "Caspar Friedrich Wolff and His 'Theoria Generationis,' 1759," *Journal of the History of Medicine* 16 (1961): 124–44; Duchesneau, "Haller et les théories de Buffon"; Peter McLaughlin, "Blumenbach und der Bildungstrieb: Zum Verhältnis von epigenetischer Embryologie und typologischem Artbegriff," *Medizinhistorisches Journal* 17 (1982): 357–72; Mocek, "Caspar Friedrich Wolffs Epigenesis-Konzept"; Ilse Jahn, "Georg Forsters Lehrkonzeption für eine 'Allgemeine Naturgeschichte' (1786–1793) und seine Auseinandersetzung mit Caspar Friedrich Wolffs 'Epigenesis'-Theorie," *Biologisches Zentralblatt* 114 (1995): 200–206.

25. It is important to note that in the 1780–81 versions dealing with that concept, Blumenbach avoided the term *epigenesis* and consistently sought to discriminate his idea from that of Wolff. Sloan, "Preforming the Categories: Kant and Eighteenth-Century Generation Theory," *Journal of the History of Philosophy* 40 (2002): 229–53, notes Blumenbach's aversion to the term *epigenesis* in his early texts on the *Bildungstrieb*.

26. Genova, "Kant's Epigenesis of Pure Reason," *Kant-Studien* 65 (1974): 269.

27. In his influential challenge to C. F. Wolff, Haller hit upon this: "why does this '*vis essentialis*,' which is one only, form always and in the same places the parts of an animal which are so different, and always upon the same model, if inorganic matter is susceptible of changes and is capable of taking all sorts of forms?" (cited in Aulie, "Caspar Friedrich Wolff," 140).

28. That did not escape younger intellectuals in Germany who did keep abreast of the latest developments in natural science and who could sense in Kant's work, even of the later 1780s, a position that did not quite incorporate the then current level of scholarship. In the aftermath of the publication of the third *Critique*, young philosophers, steeped in the latest science, came to find his posture insupportable. Hence here was one of the impulses that led to German Idealism.

29. Kant, *KU*, Ak. 5, 394.

30. Kant, *Metaphysische Anfangsgründe der Naturwissenschaft*, Ak. 4, 544.

31. Hence Kant situated himself squarely in the tradition of the new scientific rationalism. For an old but still trenchant assessment of this view, see E. A. Burtt, *Metaphysical Foundations of Modern Physical Science* (Garden City: Doubleday, 1954). For a more recent, penetrating analysis, see Gerd Buchdahl, *Metaphysics and the Philosophy of Science. The Classical Origins: Descartes to Kant* (Cambridge, MA: MIT Press, 1969).

32. For a recent study of Kant's theory of organic form, see Reinhard Löw, *Philosophie des Lebendigen: Der Begriff des Organischen bei Kant, sein Grund und seine*

Aktualität (Frankfurt, 1980), esp. 138ff. For the older literature, see P. Menzer, *Kants Theorie von der Entwicklung* (Berlin, 1911); K. Roretz, *Zur Analyse von Kants Philosophie des Organischen* (Vienna, 1922), 112–50; Emil Ungerer, *Die Teleologie Kants und ihre Bedeutung für die Logik der Biologie* (Berlin, 1922), 64–132; Paul Bommersheim, "Der vierfache Sinn der inneren Zweckmäßigkeit in Kants Philosophie des Organischen," *Kant-Studien* 32 (1927): 290–309; and Hans Jorg Lieber, "Kants Philosophie des Organischen und die Biologie seiner Zeit," *Philosophia Naturalis* 1 (1950): 553–70.

33. Kant, *KU*, Ak. 5, 400, 405–6.

34. Kant, Preface to *Metaphysische Anfangsgründe der Naturwissenschaft*, Ak. 4, 468, 470–72.

35. *KU*, Ak. 5, 374–75.

36. *KU*, 311.

37. Ibid. While our reason is driven that far, it cannot make the final step and establish "the determinate concept of that supreme intelligence." Kant concluded: "the concept of a deity, which would be adequate for our teleological judging of nature, can never be derived according to mere teleological principles of the use of reason (on which physicotheology alone is based)" (*KU*, 440). The notion of a transcendent, intelligent cause can be promoted from its heuristic, theoretical use to the full-fledged notion of God only through "ethico-theology," Kant argued in the closing segments of the third *Critique*. Yet the kind of being physico-theology required to make the world coherent for discursive understanding tallied with the kind of being practical reason required in terms of the indubitability of the moral law and all the consequences it brought in its train. This, in turn, allowed Kant to translate his "theist" notion—merely conjecturally, of course—from the one sphere to the other. The resultant notion of "Providence" tallied well with traditional religion.

38. Kant, *Der einzig mögliche Beweisgrund zu einer Demonstration des Daseins Gottes*, Ak. 2, 63–164, citing 114–15.

39. Sloan's "Preforming the Categories," 235–36, has made this perfectly clear.

40. Shirley Roe, *Matter, Life and Generation: Eighteenth-Century Embryology and the Haller-Wolff Debate* (Cambridge: Cambridge University Press, 1981).

41. Zöller, "Kant on the Generation of Metaphysical Knowledge," in *Kant: Analysen—Probleme—Kritik*, ed. H. Oberer and G. Seel (Würzburg: Königshausen and Neumann, 1988), 79.

42. Haller and the incipient Göttingen school acknowledged such an idea in principle but could not bring themselves to accept in its Buffonian formulation (Sloan, "Buffon, German Biology," 122–23; Lenoir, "The Göttingen School," 120–23).

43. Kant, *Critique of Pure Reason*, 66, emphasis added.

44. Sloan, "Preforming the Categories," 230–31.

45. Kant, *Critique of Pure Reason*, B145–46.

46. See Kant, *Prolegomena zu einer jeden künftigen Metaphysik, die als Wissenschaft wird auftreten können*, Ak. 4, 253–384, citing 319.

47. Zöller, "Kant on the Generation," 80–84, discusses uses of epigenesis in Kant's lectures and *Reflexionen*, but there is no reason to suspect that any of these date significantly to before 1786.

48.   Kant, "Recensionen von J. G. Herders *Ideen zur Philosophie der Geschichte der Menschheit Theil 1. 2.*," Ak. 8, 62–63.

49.   Kant, "Bestimmung des Begriffs einer Menschenrace," (1785) Ak. 8, 97; translated in Lovejoy, "Kant and Evolution," in *Forerunners of Darwin, 1745–1859*, ed. Bentley Glass et al. (Baltimore: Johns Hopkins University Press, 1959), 184.

50.   *KU*, 400, 409.

51.   *KU*, 435.

52.   Kant, "Über den Gebrauch teleologischer Principien in der Philosophie," Ak. 8, 157–84, citing 179.

53.   Zöller, "Kant on the Generation," 81.

54.   "In chemistry one distinguishes between matter *tanquam* **eductum** (e.g., oxidized potassium [*Potassche Aschensalz*] is an educt)—what was there before has only taken on a new form, [and] *tanquam* **productum**, of which there was nothing there before. . . . The systems of human generation are 1) *involutionis* (encapsulation [*Einschachtelung*]), 2) epigenesis, [the claim] that humans are produced entirely anew. In the first case man is an educt, in the second a product; if we have grounds for accepting the system of epigenesis, then we should assume man is a product,—*propagatio per traducem* would then transpire in the case of souls. Is it possible that the soul could produce other substances?—This is contrary to first principles, for substances persist—and they would in that case have to be composite—and the soul is a simple substance. The claim for a *propagatio per traducem* is absurd [*ungereimt*] and has not the least concept of possibility" (Kant, *Vorlesungen über Metaphysik* Ak. 28, 684). The same ground is covered in slightly different language in another set of notes from the metaphysics lectures: "A substance [*Materie*] is 1) an educt, that is what was once in another substance but is now presented separately[; or] 2) A product, what before was not yet present, but now is generated [*erzeugt*] for the first time. . . . Whoever assumes the soul is an educt . . . assumes the system of the preexistence of souls. Whoever assumes the soul is the product of the parents believes in the system of propagation. . . . The systems of human generation are twofold: 1) involution (encapsulation [*Einschachtelung*]): all children lay within their original parents, 2) epigenesis, according to which humans, as far as bodies are concerned, are brought forth entirely new. According to the first, man is a mere educt (educt was already present before birth, only in combination with other material, so that it appears by disaggregation). If we have cause to assume the system of epigenesis, then we also have cause to assume the soul as a product, because otherwise the soul must have existed somewhere, and then become conjoined with the newly created body. Thus here one would have to assume in connection with the soul a *propagatio per traducem*. But a substance cannot generate another substance, and the same is true for the soul. A soul cannot put forth other souls from itself, for then it would be a composite. . . . To assume the propagation of human souls *per traducem* is absurd, because we do not know how to judge it at all. If the soul were a product, then the souls of the parents would have a creative force [*schöpfende Kraft*]. All generation of a substance is *productio ex nihilo*, creation; because before the substance there was nothing. A creature, however, does not itself have a creative, but only a developmental [*bildende*] force, i.e., [the ability] to divide or to compound things that are already given. There is no alternative, accordingly, but to assume the soul is preformed [*präformiert*], however it may be with the creation of the body" (Kant, *Vorlesungen über Metaphysik*, Ak. 29, 760–61).

55.   It reappears in a crucial context: Kant, *KU*, 423.

56.   Zöller makes the point that Kant distinguished in his *Reflexionen* between an *epigenesis psychologica* and an *epigenesis intellectualis*, and it is really the latter, the origin of the categories, that is at issue at B167. That is altogether correct, yet what concerns me here is not the question of the origin of the soul (a transcendent metaphysical concern, according to orthodox interpretation of the critical philosophy), but rather the way in which Kant conceived of preformation and epigenesis (Zöller, "Kant and the Generation," 82–83).

57.   Kant, *Critique of Pure Reason*, B167. See Julius Wubnig, "The Epigenesis of Pure Reason," *Kant-Studien* 60 (1968): 147–52; Genova, "Kant's Epigenesis"; Zöller, "Kant and the Generation"; H. Ingensiep, "Die biologischen Analogien und die erkenntnistheoretischen Alternativen in Kants Kritik der reinen Vernunft B §27," *Kant-Studien* 85 (1994): 381–93; and Sloan, "Preforming the Categories."

58.   Ingensiep, "Die biologischen Analogien," 385.

59.   Ibid. There remains, even for Ingensiep, however, a clear structural correlation: "As, according to the epigenesis theory, unformed inorganic matter gets transformed under the direction of a "purposive endowment" into something entirely new via the *Bildungstrieb* and an organism is produced, so via the categories and the raw material of sensibility empirical knowledge is 'produced.' The organizing productivity, however, lies entirely on the side of categorizing understanding. As, via the requisites of the epigenesis theory, from something unformed and unpurposive gradually something specially formed, purposive (according to the most inward interfusion of generative fluids) gets produced, so, similarly, one can conceive the unifying ordering of the manifold by the categories" ("Die biologischen Analogien," 387).

60.   "[A] necessary agreement of the principles of possible experience with the laws of the possibility of nature can only proceed from one of two reasons: either these laws are drawn from nature by means of experience, or conversely nature is derived from the possibility of experience in general and is quite the same as the mere universal conformity to law of the latter. The former is self-contradictory, for the universal laws of nature can and must be cognized a priori (that is, independent of all experience) and must be the foundation of all empirical use of the understanding; the latter alternative therefore alone remains" (Kant, *Prolegomena*, Ak. 4, 319). "There are only two ways in which we can account for a *necessary* agreement of experience with the concepts of its objects: either experience makes these concepts possible or these concepts make experience possible. The former supposition does not hold in respect of the categories (nor of pure sensible intuition); for since they are—concepts, and therefore independent of experience, the ascription to them of an empirical origin would be a sort of *generatio aequivoca*. There remains, therefore, only the second supposition—a system, as it were, of the *epigenesis* of pure reason—namely, that the categories contain, on the side of the understanding, the grounds of the possibility of experience in general" (Kant, *Critique of Pure Reason*, B166–67).

61.   "Indeed, it might be argued that, had not epigenesis become popularly linked with espousal of spontaneous generation, it would have claimed majority support several decades earlier" (Bodemer, "Regeneration," 28).

62.   "A middle course may be proposed between the two above mentioned, namely, that the categories are neither *self-thought* first principles a priori of our

knowledge nor derived from experience, but subjective dispositions [*Anlagen*] of thought, implanted in us from the first moment of our existence, and so ordered by our Creator that their employment is in complete harmony with the laws of nature in accordance with which experience proceeds—a kind of *preformation-system* of pure reason" (Kant, *Critique of Pure Reason*, B167–68).

63. *Critique of Pure Reason*, B167–68.

64. Genova, "Kant's Epigenesis," 269; Günther Zöller, "From Innate to a priori: Kant's Radical Transformation of a Cartesian-Leibnizian Legacy," *Monist* 72 (1989): 227.

65. Sloan, "Preforming the Categories," 242, recognizes the decisive significance of this.

66. Blumenbach, *Über den Bildungstrieb und das Zeugungsgeschäfte* (1781; reprint: Stuttgart: G. Fischer, 1971); Sloan, "Preforming the Categories," 247.

67. Kant, "Über den Gebrauch teleologischer Prinzipien," Ak. 8, 180n.

68. Kant did not own the first version of Blumenbach's book, but ownership is a poor index of Kant's voracious reading. It is not at all impossible that Kant should have read it or read about it. Blumenbach was clearly a celebrity in the medical science of Germany in his day, so that his work would very likely have appeared in bookshops in Königsberg for the attention of the medical faculty of that university. There would also have been reviews of the work, which Kant may have encountered. In addition, there were two Latin versions of Blumenbach's work, one published in 1785 and a second in 1787. Kant did use the German term, and therefore it is likely that he was aware of the German version of Blumenbach's theory.

69. Kant, *KU*, 419n. Forster had introduced something like this in his "Noch etwas über die Menschenraßen," *Teutsche Merkur* (1786): 57–86, 150–66.

70. Kant, "Über den Gebrauch teleologischer Prinzipien," Ak. 8, 180n.

71. Blumenbach, *Handbuch der Naturgeschichte*, 4th ed. (Göttingen: Dieterich, 1791), 13. But: "In fact, the *only clear substantive difference* in the key formulations of the theory of the *Bildungstrieb* between the 'more mature' and the 'immature' phase is the replacement of an 'innate' drive by a 'general' drive" (McLaughlin, "Blumenbach und der Bildungstrieb," 371).

72. "The term *Bildungstrieb* just like all other life forces such as sensibility and irritability explains nothing itself, rather it is intended to designate a particular force whose constant effect is to be recognized from the phenomena of experience, but whose cause, just like the causes of all other universally recognized forces, remains for us an occult quality. That does not hinder us in any way whatsoever, however, from attempting to investigate the effects of this force through empirical observations and to bring them under general laws" (Blumenbach, *Handbuch der Naturgeschichte*, 2nd ed. (Göttingen: Dieterich, 1782), 19).

73. Blumenbach, *Über den Bildungstrieb* (Göttingen: Dieterich,1789), 71.

74. Kant to Blumenbach, August 5, 1790, Ak. 11, 176–77.

75. *KU*, 424.

76. Sloan, "Preforming the Categories," 248.

77. *KU*, 423.

78. Ibid.

79. *KU*, 424.

80. Genova, "Kant's Epigenesis," 465; Zöller, "Kant and the Generation," 90.

81.   Sloan, "Preforming the Categories," 252–53.

82.   Clark Zumbach, *The Transcendent Science: Kant's Conception of Biological Methodology* (Den Haagen: Nijhoff, 1984), 99.

83.   "[T]his claim has a decidedly negative import; it is essentially just an affirmation that the mechanical conception of nature and its conception of causality fails to provide a complete characterization of living systems. . . . Thus, the claim that there are free causes in living systems has no ontological force. It is rather a transcendental claim, i.e., one concerning the possibility of our judgments" (Zumbach, *The Transcendent Science*, 107).

84.   *KU*, 405–10.

85.   Robert E. Butts, "Teleology and Scientific Method in Kant's *Critique of Judgment*," *Nous* 24 (1990): 5.

86.   "Teleology and Scientific Method," 7; Peter McLaughlin, *Kant's Critique of Teleology in Biological Explanation: Antinomy and Teleology* (Lewiston: E. Mellen Press, 1990).

87.   And thus his effort to "police" the practices of the experimental physics of his day was unavailing. See my "'Method' versus 'Manner.'"

88.   Kant, *Metaphysische Anfangsgründe der Naturwissenschaft*, Ak. 4, 467–69 and passim.

89.   "Whilst the extensionalist mathematical Newtonian approach offers the potential for (mathematical) a priori processing of physical nature, the price which this pays is that since forces do not have in this scheme any basic or 'essential' place, they have (because of the conceptual doubt attaching to them) to be introduced ad hoc (from 'without'), by way of hypothesis only. The objection to this, of course . . . , [is]—that such a basic and powerful notion as force (let alone the force of attraction) ought not to be surrounded with the suspicion which—particularly during the seventeenth and eighteenth centuries—surrounded anything 'hypothetical' in science" (Gerd Buchdahl, "Kant's 'Special Metaphysics' and the *Metaphysical Foundations of Natural Science*," in *Kant's Philosophy of Physical Science*, ed. Robert E. Butts (Dordrecht: Reidel, 1986), 150–51).

90.   Lovejoy, "Kant and Evolution," 185.

91.   See Lewis Beck, "Lovejoy as a Critic of Kant," in *Essays on Kant and Hume* (New Haven: Yale University Press, 1978), 61–79.

92.   Frederick Gregory, "Kant's Influence on Natural Scientists in the German Romantic Period," in *New Trends in the History of Science*, ed. Robert Paul W. Visser et al. (Amsterdam: Rodopi, 1989), 57.

93.   Goethe, cited in Jane Oppenheimer, *Essays in History of Embryology* (Cambridge, MA: MIT Press, 1967), 136.

3

Reflexive Judgment and Wolffian Embryology

Kant's Shift between the First and the Third Critiques

Philippe Huneman

Abstract

The problem of generation has been, for Kant scholars, a kind of test of Kant's successive concepts of finality. Although he deplores the absence of a naturalistic account of purposiveness (and hence of reproduction) in his pre-critical writings, in the First *Critique* he nevertheless presents a "reductionist" view of finality in the Transcendental Dialectic's Appendices. This finality can be used only as a language, extended to the whole of nature, but it must be filled with mechanistic explanations. Therefore, in 1781, mechanism and teleology are synonymous languages. Despite the differences between its two authors, the Wolffian embryology, exposed in the *Theorie der Generation* (1764) and debated by Blumenbach's dissertation on *Bildungstrieb*, enabled Kant to resolve the philosophical problem of natural generation, and subsequently to determine what is proper to the explanation of living processes. Thus, in the Third *Critique* he could give another account of purposiveness, restricted to the organism and more realist than his former one; this philosophical reappraisal of purposiveness in embryology required the new concept of "simply reflexive judgment" and the correlated notion of "regulative principle." Thus framed, this naturalized teleology provided some answers to the Kantian problem of order and contingency after the end of classical (Leibnizian) metaphysics.

Introduction

The First and the Third *Critiques* deal with purposiveness.[*] Although sometimes convergent, their accounts differ slightly. Here I want to clarify the

---

[*] I thank Robert Richards for his constructive comments and critiques on the first version of this text. The English language has been corrected by Cecilia Watson of the Fishbein Center at the University of Chicago.

need for a reassessment of purposiveness in the *Critique of Judgment*, and the changing meaning of this concept. I argue that one central motive for this shift is the embryological question and the epigenetist theory, and I show that this problem is longstanding in Kant's thinking. But the issue raised by this problem and its eventual solution deeply concerns some major conceptual problems, which stem from Kant's criticism of Leibnizian metaphysics.

I begin by giving the background of the concept of purpose in the dialectics of the First *Critique*, which can be traced back to some of the precritical writings. I then survey the debates in the field of embryological theory and point out the major epistemological novelties. This will provide the key concepts needed to understand Kant's shift in the Third *Critique* and the concept of organism therein. It will primarily explain the reason why the Third *Critique* correlates the concept of purpose with the faculty of reflexive judgment, whereas in the Transcendental Dialectics of the First *Critique* this concept was supposed to belong to reason.[1] Finally, I stress the consequences of this shift for the critique of metaphysical concepts.

1. The Doctrine of Finality in the Transcendental Dialectics and Its Background

In *Der einzig mögliche Beweisgrund zu einer Demonstration des Daseins Gottes* (1763), Kant dealt with the concept of order. The same speculations had laid the foundations for the scientific attempt he had made some years before, in the *Allgemeine Naturgeschichte und Theorie des Himmels* (1755). There, his method was to simulate the genesis of an apparent order by the purely natural lasting action of the laws of movement. He sees "an ordered and beautiful whole developing naturally from all the decomposition and dispersion (of the matter). This whole isn't produced by chance and in some contingent manner, but we notice that some natural properties necessarily generate it."[2] He thought that we, unlike Newton, could thus naturally explain some harmonious features of the solar system.[3]

In the *Beweisgrund*, he distinguished between *necessary* and *contingent* order. When there is order, we perceive some finality, because the parts seem to play a role in the whole. Sometimes, a lot of useful features are the collective effect of a few laws of nature, though no part had been intended for a particular role. This is especially true of the winds on the seashore, which seem to serve a human purpose, when in reality their course is determined by general motions in the atmosphere. The apparent utility of a thing (and, especially, its apparent utility for man) stems, in fact, from its natural behavior according to laws and its relationships to other things. This is the necessary order, or the purposiveness conceived as a *system*. In it, all the laws of nature involved do share the same necessary unity.[4] Hence, the system necessarily develops according to the laws of nature. When there is no such system, but there is nonetheless some finality, then the parts receive their purpose from some intention: this finality is a *technical* one, and the order is said to be *contingent* because the parts could have not been there all at once.

More important, such a whole is constituted by the action of various laws that operate independently of one another and that therefore must be unified by some external constraint: "The principle of a kind of operation according to a given law, is not at the same time the principle of another operation in the same being according to a law."[5] One principal example of this contingent order is the living body: "Creatures of the animal and vegetal kingdoms present, in many ways, the most amazing examples of a contingent unity, a unity attributed to a deep wisdom. Here are channels guiding the sap, others propelling the air, others which push it up, etc.—a big manifold, any single part of which has no ability to fulfil the functions of the others, and within which the convergence of each toward a common perfection is artistic [*künstlich*], so that the plant, with its relations to so many various purposes, constitutes an arbitrary contingent unity."[6]

However, most of the apparently technical features can be reduced to necessary order, allowing us to avoid direct reference to a transcendent maker: "theologians consider any perfection, any beauty, any natural harmony as an effect of the divine wisdom, whereas a lot of those follow with a necessary unity from the most essential laws of nature."[7] The epistemology of the *Himmelstheorie* precisely reduces contingent to necessary order. The *Beweisgrund* infers from this epistemology a theological methodology: "One will seek the causes, even of a most advantageous disposition, in the universal laws, which, excepting any accessory consequences, concur with a necessary unity in the production of those effects."[8] This is a methodology of the "unity of nature,"[9] and in many respects, the Transcendental Dialectics will elaborate on this equation of God with the unity of nature. Therefore, Kant is not rebutting theology. In fact, this new method provides a new proof of God, depicting Him no longer as a Great Maker of things—according to a technical schema—but as the foundation of the essential unity of laws, which cannot be understood and conveyed in the technical vocabulary.[10]

Kant hesitates when considering whether contingent order cannot be reduced to the necessary effect of natural laws. Although living creatures seem to be a good example of this irreducibility, he cautions that many of the technical features of a living creature could in fact emerge from natural systems: "Not only in the inorganic nature, but even in the organized nature, one will suppose a much greater role for necessary unity than appears at first glance. Even in the building [*Bau*] of an animal, we are allowed to suppose that a simple disposition [*Anlage*] is given a wide capability to produce many useful consequences, consequences to the generation of which we would prima facie think necessary to appeal to many particular dispositions."[11]

Kant was thus led to the problem of *generation*. Generation seems to be a major objection to the reducibility of contingent orders to necessary orders: one cannot imagine a system that could account for the inheritance of forms.[12] However, the doctrine of preformationism (founding all generations on a technical order) seems methodologically as well as philosophically wrong regarding the epistemological requirement to maximize the reduction of contingent orders into systematic orders:[13] "The philosophical dimension [*das Philosophische*, which here means "rational"] of this way of thinking is even slighter when each plant or animal is immediately subordinated to the [divine] creation, than when, with the exception of some immediate creations, the other [living] products are subordinated to a law of the reproductive capability (and not of the simple power of development), because in this last case nature is much better explained according to an order."[14] But on the other hand, epigenetic theories of generation, such as Buffon's "*molécules organiques*" and "*moules intérieurs*" or Maupertuis's *Venus physique*, are not satisfying, because, even if they fulfill the epistemological requisite, they do not provide any intelligible account of generative mechanisms: "The internal molds [*innerliche Formen*] of M. De Buffon, the elements of organic matter which, according to M. De Maupertuis, combine together according to the laws of desire and repulsion, are at least as inexplicable as the phenomenon itself, if not wholly arbitrarily imagined."[15] The question of generation at this period seems crucial to finding the limits of what we could call the genetic epistemology (i.e., to reduce technical orders to systems), but at the same time it seems scientifically aporetic. We can call this the *generation dilemma*.

The concept of purpose in the Transcendental Dialectic must be understood in the context of this dilemma. In the *Critique of Pure Reason*, the lawlikeness of nature is constituted by our understanding, according to the conditions of a possible experience. Since a lawlike but orderless nature is possible, the *order of nature* is not guaranteed by this constitution, but some rational presuppositions, such as the hierarchic classification of things into genera and subgenera, do yield such an order.[16] Compared with the *Beweisgrund*, this means that the difference between system (= necessary order) and technique (= contingent order) has been replaced by the difference between *nature* (lawlike) and *order of nature* (constituted by the transcendental presuppositions of reason). But this requires that the difference between necessity and contigency is no longer relevant here; the order of nature is necessitated by a kind of necessity other than that of nature itself with its laws (grounded on the principles of the synthetic judgments and the principles of the metaphysical foundations of the science of nature). Such a difference was emphasized by Kant when he opposed the constitutive principles of the understanding and the so-called regulative principles of reason.[17]

If we want, we can speak of God harmoniously creating the world, but that is *mere language* that we employ in order to express knowledge in a *systematic* way. Reason confers systematicity to nature; and this systematic unity appears as *purposiveness:* "the highest formal unity, resting on the sole principles of reason, is the purposive unity" (A686/B714). Kant thinks that the idea of a system presupposes a kind of plan, an architecture,[18] which implies a design, and hence a purpose.[19] Here, the concept of purpose is related only to systematicity (leaving aside this other part of purposiveness that the *Beweisgrund* named "technique"). The *Critique of Pure Reason* follows to its end the reductionist trend of the *Beweisgrund:* the contingent order, or finality as technique, is explained away. Finality is just the *language of reason in front of nature*, language by which it settles its correlate, namely, the *order* of nature. To make possible a science investigating nature according to its universal laws, reason must presuppose order and systematicity, and the language of systematicity is purposiveness. Finality is a mere language because it expresses systematicity only as a necessary presupposition of reason pertaining to the science of nature as science of mechanical universal laws. That is the meaning of the "as if" in the following sentence: "For the regulative laws of systematic unity would have us study nature as if systematic *and purposive* unity together with the greatest possible manifoldness were to be encountered everywhere to infinity" (A701/B729; emphasis added).

This very definition of purposiveness preserves it from any theological commitment by binding it strictly to the requirements of *Naturforchung,* namely, the consideration of the universal laws of nature and the rational ordering of them: "since, however, the principles that you are thinking of have no other aim than to seek out the necessary and greatest possible unity of nature, we will have the idea of a highest being to thank for this so far as we can reach it, but we can never get around the universal laws of nature so as to regard this purposiveness of nature as contingent and hyperphysical in its origin, without contradicting ourselves, *since it was only by taking these laws as our aim that things were grounded of the idea;* hence we won't be able to consider this purposiveness as contingent and hyperphysical by its origin, because we were not justified in admitting above nature a being with all the qualities here supposed, but we were only entitled to put at its basis the Idea of such a being in order to consider, following the analogy of the causal determination, the phenomena as systematically linked together" (A700/B728; emphasis added). I emphasized the passage that firmly binds finality to the needs of systematizing universal laws, which makes it a mere language made to express the interests of speculative reason. This focus on universal laws of nature, as allowing and requiring us to talk in a purposive language, contrasts with Kant's concern in the Introduction to the Third *Critique,* where the emphasis will be put on the empirical laws of nature and their compatibility (so that we are entitled to forge empirical concepts).

There is no point in asking about the reality of purposiveness, because that would be to ask about the reality of a language. Therefore, there is no contingent (and hence technical) order, because a mere way of describing things does not involve any real creator or intention, hence no technique.[20] Because of this merely linguistic sense of finality—and this point is especially noteworthy—no counterexample, namely, no discovery of any mechanistic bond where a purposive one (*nexus finalis*) was expected, could invalidate this purpose-language (A687/B715). There is nothing empirical in finality; thus, after the *Beweisgrund*, this use of purposiveness is preserved from Kant's criticism of those theologians eager to pick out particular purposes in nature.

Hence, finality does not *explain* anything. That is why we can freely use the purpose-language. It is intended to access only the best and the most extended mechanistical explanations in terms of universal laws of nature. That is why pointing to a supposed purpose as the explanation of some natural feature is totally misleading and commits one to the sophism of the "lazy reason": believing that this saves us an explanation, whereas it is precisely the starting point of the explanation.[21] The example of the physiologist illustrated this point: when he says that this organ has a function, he is then required to elucidate the natural ways according to which the function is fulfilled; else, by conceiving his function ascribtion as an explanation, he would be guilty of "lazy reason."[22]

The "Aim of a natural dialectic" formulates the point quite clearly: "It must be equivalent for you, wherever you perceive the purposive and systematic unity of nature, to say: God designed it wisely, or nature wisely arranged [*geordnet*] it. Because the greatest purposive and systematic unity, which your reason demands to use as a regulative principle at the outset of any investigation into nature, was precisely what allowed us to put at the beginning the idea of a supreme intelligence, as a result of the regulative principle. Now, the more you find finality in nature according to this principle, the more you see your idea confirmed."[23] "Natural arrangements" (therefore naturally, namely, *physically*, produced) and "divine will" are equivalent in meaning, because the divine intelligence is the scheme of—that is, the best language with which to formulate—the systematic unity, which can be seen as the ordering of nature *through its immanent* laws. "God" and "nature" are two ways of applying the same presupposition of the systematic unity of nature, which concerns only the laws of nature and, therefore, the possibility of nomothetic explanation. The physical explanation according to laws, and the transcendent account using God's intentions, prove themselves synonymous from a critical point of view, of course, even if the latter has less explanatory power. Since finality does not explain anything, identifying purposive arrangements means only gesturing toward future research into mechanical connections. But Kant is not a Spinozist; he does not see purposiveness as illusory, even as a useful illusion: finality is a *justified* language, because it presents the order of nature (and not nature itself), which is a requisite of *reason*. To this extent, purposiveness and mechanism are *not opposites*, as they will be sub-sequently in the *Critique of Judgment*. No antinomy stands between those two instances; on the contrary, they are quite synonymous.[24]

If technical finality is ruled out, the living creatures see their status quite obscured when compared with the *Beweisgrund*. Of course, as the purpose-lan-guage is justified, Kant can, as we have seen, recognize the validity of some classical physiological principles, such as "all in an animal has its utility."[25] But this example comes after an illustration about the North Pole's contrib-uting to the Earth's stability by its own flatness. Indeed, a living being is not *essentially* purposive, it is *paradigmatically* purposive—and this is somehow cor-relative to the requirement that the presupposition of purpose should be as general as possible in order to avoid the sophism of "lazy reason." Finality is the language with which physiologists and anatomists can extend their knowl-edge of animals, because in those sciences there are lots of opportunities to use this language—more opportunities than in many other sciences. If we compare the first *Critique's* page on this principle with the *Lectures on the Philosophy of Religion*, given a few years after, we see that the same sentence shifted to "the presupposition that all, *in the world*, has its utility," while the *Critique's* text said "the presupposition that all, *in the animal*, has its utility."[26] The fact that Kant can shift freely from one meaning to the other indicates that purposiveness, as a language, is not restricted to one ontological type. And in this notion of pur-posiveness as language, the concept of utility is conspicuous. Thus, for Kant, in 1781, purposiveness implies a rational language using utility considerations as a *Leitfaden*, and mainly exemplified by living beings. A consequence worth emphasizing is that the difference that structures the Third *Critique* is the dif-ference between a formal concept of purposiveness as ordering of empirical laws in the Introduction[27] and an objective concept of purposiveness con-cerned with the production of singular entities; and this difference is necessar-ily absent from the theory of purpose in the First *Critique*, where purposiveness is, by nature, addressing the whole of nature in its form.

Later the Third *Critique* would seem to reinforce the ontological commit-ments of finality, by focusing on the organisms as *Naturzwecke*, which will be purposive *objects* rather than a mere language of purposiveness, of which living beings would be the privileged examples of its application. Therefore, such a change in the concept must be referred to a modified perspective con-cerning the living beings.

2. The Peculiar Epistemological Features of Wolffian Embryology, and the Vital Forces

Kant developed his theories of generation further, beginning with his 1775 work on the human races "*Von den verschiedenen Menschenrassen.*" But we cannot take for granted that the *Critique* presents a position on those questions that is wholly compatible with these writings, because they invoke germs and dispositions (*Keime* and *Anlagen*)[28] at the origin of growing beings, which could be activated in the appropriate circumstances by contact with the milieu.[29] Hence, this theory uses a kind of teleological preadaptation, which does not enter into the *Critique's* conceptual frame. Dispositions are localized in the *Stamm* of the species or the subspecies, which means that they are transmitted in the reproductive power: they are hereditary. They cannot be deduced from those same physical laws that describe the environment, because they prepare the organism to adapt to many possible environments.[30] Hence we cannot apply here the synonymy of the language of purposiveness and the mechanical description established in the *Critique of Pure Reason*. The theories of the essays on races and their conceptions of preadaptation, considered retrospectively, seem to invalidate any identification of finality with systematicity (as a scheme for extending knowledge along universal nomothetical guidelines), and require us to work through the previous category of the technique of nature.[31]

In fact, after the *Beweisgrund*, Kant found a way to escape the generation dilemma he had articulated in that text. He could embrace an epigenetic theory that, unlike Maupertuis's and Buffon's, fulfilled the scientific requisites. Although Kant often refers to Blumenbach, to his notion of *Bildungstrieb*, and mainly to his reluctance to derive organic devices from mere matter,[32] the scientific bases for such a solution had been laid by Caspar Wolff in his *Theorie der Generation* (1764, 1st Latin edition *Theoria generationis*, 1758), of which Kant became aware some years later. Let us settle this major point.

The later embryologists Johann Friedrich Blumenbach and Christian Heinrich Pander recognized the epistemological novelty of this work before the great Karl Ernst Von Baer, who considered Wolff's *Theoria generationis* the major achievement in the field. Previously, Maupertuis and Buffon had subscribed to a Newtonian model: the scientist aims to state the mathematical laws, which rule the forces ("penetrating forces," said Buffon) building the embryo, and such forces had to be analogous to Newtonian attraction, but should be endowed with specific qualities, and mostly a kind of selectivity. The failure of scientists to achieve this ideal led them to metaphoric formulations: they ascribed "*désir*" or "*mémoire*" to molecules. Whereas those authors, having equated generation with nutrition, define genesis by continuous growth, Wolff conceives of genesis as a discontinuous process. The embryologist's primary task is to restore the phases of this process and to state their compatibilities and incompatibilities. The mechanism involved in the process—whether it is chemical, mechanical, or spiritual—is not a matter for the scientist's consideration. "One can assert," said Wolff, "that the parts of the body have not always existed, but they form themselves little by little. . . . I do not contend that epigenesis occurs by a mixing of molecules [Maupertuis], by a kind of fermentation, or by the activity of the soul [Stahl]; I only say that it occurs."[33] Wolff therefore *changed the scientific questions.*[34] We must expose some fundamental relationships between temporal sequences (gastrulation, etc.); specifically, we must uncover their succession and their antagonisms. This method provides us with an argument against preformationism, very similar to the one used later by Von Baer in his *Entwicklungsgeschichte der Thiere* (1828) and very distant from the considerations by Maupertuis or Buffon. The strength of the preformationist position was the invocation of preexisting invisible structures. But Wolff shows that there cannot be such structures, because where we expected to see nothing we see *something*, and what we observe is thus incompatible with the hypothetical invisible structures. Wolff therefore rejected Haller's main argument for preformationism, namely, the idea that even if the preformed heart, the first organ to develop, is not seen, it is only because it stays invisible to us.[35] He replies to Haller: "We see that some parts of the body, for example the breast, at a given moment, do not exist, cannot yet exist. It is not because we did not observe the breast that we conclude it is nonexistent; it is because, where the thorax should have developed, we see the amnios in formation, which represents, as we know, the inferior prolongation of the trunk in the fœtus. Hence we conclude that the imperceptible thorax cannot exist, and therefore does not exist."[36]

Although he did not formulate the germ-layer theory, Wolff settled the framework of Pander's 1817 discovery.[37] To Wolff, embryological structures appear as folds, which then curve on themselves to produce tubes. This folding forms a blurred zone, where, by the same repeated foldings, and through sutures, the organ will build itself.[38] With these ideas of *locality* of operations and formation *by repeated simple operations*, Pander later deduces the idea of the germ-layer.

After describing the formation of the embryological structures, Wolff posits a *vis essentialis* necessary to accomplish the epigenetic process. This force is related to the crucial concept of incompatibility of the phases. Traditionally, the heart was seen as the first completed organ, and the force necessary to drive the process was ascribed to the preformed heart. But if we assume that there is no preformed organ—hence, no heart—then there must be a force located in the whole epigenetic body. "Since the heart doesn't exist yet . . . , we see that it is impossible that this motion of fluids, through which vessels are formed, be attributed to the heart. It seems clear, here, that a force is acting—a force that is similar to the force that operates in plants, and which I therefore name with the same name."[39] The *vis essentialis* had already been proved the cause of the process in plants.

Essential force in living beings stems from the specific features of the embryological field. Hence, this force has special status. It is not a natural,

Newtonian-style force, as was Haller's *vis insita*, which he believed laid in the muscles and accounted for irritability. It is not the cause of some regular effects accorded by a mathematical law. Why? The embryological theory *sets* the discontinuous phases seen by the observer *into series and order*. In effect, this discontinuity requires a *principle of continuity*, which guarantees that it is the same development, since it is not the development of a previously present structure. The *vis essentialis* provides such a guarantee; it is therefore an *epistemological requisite* for descriptive embryology. It allows us to speak of a continuous time, although the described phenomena are temporally incompatible. This is a fundamental departure from Buffon's or Maupertuis's epigenetic forces, which were plainly conceived on the Newtonian model, and then endowed of some determinate ontological nature. Here, force means the methodological reverse of the order of the layers' appearance and their rolling.

The relationship between time and force is therefore radically new. The embryological time is not the empirical time of the ordinary discourse, nor the time of physics, indexed by the variable t; it is a time reconstructed as an order founded on a set of compatibilities, incompatibilities, and conditioning relationships between transitory formations. This entails a crude characterization of the difference between preformationism and epigenesis: to the former, life is a form (hence, time is external to its essence); to the latter, life is a force (hence, time is the very condition of its manifestation). If this is the fundamental characteristic of epigenesis, shared by Wolff and Blumenbach notwithstanding their theoretical differences, then when we read the Third *Critique* we can understand why Kant was committed to epigenesis despite his reservations about the full-fledged theory of epigenesis.[40]

Of course, if the "theory of generation" is, as Wolff asserted, a "demonstration" of the anatomy (whose corollary is physiology), it should "give the causes of the determination and combination of parts": but here, "causes" must not be taken in a Newtonian sense, as referring to a mathematical law; rather, it should be read historically, as a retrospective reconstruction of a series' order.[41] Embryology cannot be a "physics of life," as Buffon and Lamarck believed; it cannot be continuously related to hydraulics, as in the previous Cartesian model and Lamarck's later construct.[42]

That is why the Wolffian theory forms the basis of Kant's escape from the *Beweisgrund* generation dilemma—which, in fact, was the same as Blumenbach's when he turned away from Haller's preformationism.[43] Because his own theory of heredity in the *Rassenschriften* avoided a strict preformationism, it was compatible with the Wolffian premises. More than vaguely reminiscent of Malebranche, Kant's distinction between efficient and occasional causes in those papers can be better understood if we see in this differentiation the necessity for a dual regime of causality when one deals with generation. But it is precisely this requirement that is met by Wolff's new explanatory strategy, as he sketched it in the *Theoria generationis*: there he established another level of intelligibility, distinct from the nomothetic explanation, in order to explain (i.e., describe) generation. Not until Kant's *Über den Gebrauch der teleologischen Principien in der Philosophie* (1788),[44] and then the sixth edition of Blumenbach's *Handbuch der Naturgeschichte*, is this other regime of cause termed "teleological." Indeed, when Wolff later sketches in the *Objectiorum* some theories of the *materia qualificata* to account for the constancy of hereditary forms in varieties through generation,[45] he veers close to Kant's idea of germs and dispositions enclosed in an "original organization"[46] and proper to one species.

But the major Kantian mention of epigenesis and embryological theorizing must be found in the Third *Critique*. Its concept of *Naturzweck*, which departed from the Transcendental Dialectic's doctrine of finality, and its assertion that *Organisierte Wesen* must be thought as *Naturzwecke*, rests on Kant's assessment of this epistemological turn. Let us demonstrate this, first by Kant's use of the concept of "force."

3. The Theory of Natural Purposes and Organisms in the Critique of Judgment: Kant's Shift

How did the *Critique of Judgment* conceive of the specificity of organisms?

> An organized being is thus not a mere machine: for the latter has only the moving force *[bewegende Kraft]*, while the former has a formative force *[bildende Kraft]*, of a kind that it imparts to materials not possessed of it (it organizes these materials). Therefore it has a propagative *[fortpflanzende]* power of formation that cannot be explained through the mere power of motion *[Bewegungsvermögen]* (the mechanism).[47]

So the specificity of living things, as organized beings, is specificity in the inherent force. Their abilities to reproduce, and to assimilate brute matter, exceed the forces usually proposed in Newtonian physics, because those forces can be *transmitted* (according to Leibnizian laws) but not *propagated;* but the forces in organized beings can be infused into some material in order to transform themselves into the organism itself. To be an organism is to perform a specific activity, namely, to *organize*. That is why the concept of an organized being, stated precisely in this paragraph, is a *self-organizing being*. The activity of organizing requires a force incongruous with physical forces, a "formative force." This locution refers both to Blumenbach (*Bildungstrieb*) and Wolff (because it speaks of forces, *Kräfte*, as does Wolff). But such a force shares the epistemological status of Wolff's *vis essentialis*; for example, it is an epistemological notion that is required for rendering intelligible those epigenetic features that cannot be understood through mechanical laws.[48]

Kant's main assumption in the paragraph below will help us make this point. Here Kant identifies *Naturzwecke* and organized beings. Let us consider the text:

> If something, as a natural product, has to embrace in itself and in its internal possibility some relation to purposes; that is, if the thing must be possible only as a natural purpose, and without the causality of a rational being external to it, then it must be the case, that . . . the parts of this thing are bound together in the unity of a whole, and that they are reciprocally, one to another, causes and effects of their forms. It is only in this way that it is possible that, in the reverse way (reciprocally), the idea of the whole determines the form and the binding of all the parts: not as a cause, since it would be a product of art—but as a principle of knowledge *[Erkenntnisgrund]* of the systematic unity of the binding of all the manifold contained in the given matter for the one who has to judge.

> Thus, concerning a body that has to be judged as a natural purpose in itself and according to its internal possibility, it is required that the parts of it produce themselves *[hervorbringen]* together, one from the other, in their form as much as in their binding, reciprocally, and from this causation on, produce a whole. In a body that would possess the concept-ruled causation proper to such a product, the concept could be judged as a cause of it according to a principle, and subsequently the binding of the efficient causes could be judged in the same time a binding according to the final causes.

> In such a product of nature each part, at the same time as it exists throughout all the others, is thought as existing with respect to [*um . . . willen*] the other parts and the whole, namely as instrument (organ). That is nevertheless not enough (because it could be merely an instrument of art, and represented as possible only as a purpose in general); the part is thought of as an organ *producing* the other parts (and consequently each part as producing the others reciprocally). Namely, the part cannot be any instrument of art, but only an instrument of nature, which provides the matter to all instruments (and even to those of art). It is then—and for this sole reason—that such a product, as *organized and organizing itself*, can be called a natural purpose.[49]

To assume that the whole determines the parts is not enough to distinguish natural purpose from technical purpose, because that is as true of watches as it is of animals.[50] So the second criterion delineates the intimate productivity of the natural purpose, according to which the parts produce one another, in their forms and relationships.[51] Kant argues in the following way: the idea of a whole determines the parts—that is "purpose." But if determination meant "cause," then we would have a mere technical object: the idea of its pattern caused its realization. To talk of natural purposes implies that this "idea of a whole" is no cause; hence it is only a *principle of knowledge*. So the *real* causes of the arrangement of parts must be the parts themselves, which thereby produce one another and their relations *according—for us—to the idea of the whole*. Hence the nature of this causation acting on and within the parts is plainly unknown to us, which means there is no analogue, even if we should in a regulative manner conceive the production of the whole as analogous to our technical activity (namely, a concept is at the source of the production of the object).[52]

To say that the "idea of the whole" is only a principle of knowledge is to say, in Kantian language, that it is only a *regulative* principle, because it is not, like the laws of nature, embodied and efficient in the phenomena themselves.[53] Therefore—here is the central Kantian idea—the *internal productivity* of the organisms, their *epigenetic* character, is *essentially bound* to the *specific epistemological character* of the biological knowledge (i.e. the regulative status of the idea of an organism). But in the words of the Third *Critique*, a judgment that states in itself the conditions under which we must enunciate it is called "simply *reflexive* judgment"; and it is opposed to the use of judgment in the physical sciences, wherein the particular is always determined by previously known rules, and hence called *determinative* judgment.[54] The idea of a whole, for a natural purpose, is thus clearly involved in a reflexive judgment.

Kant's position, therefore, consists in linking the assumption of the *reflexive character of the teleological judgment* with the endorsement of an *epigenetic productivity inherent to the organized beings*. If we recall the specific, unmechanical, or un-Newtonian status of the Wolffian embryological program, and mainly its *requirement* of a *vis vitalis* or *vis essentialis* as a condition of continuous embryological time—therefore, in Kantian terms, making no use of the determinative judgment—we see that the main assertion of the Third *Critique*, equating "natural purpose" and "organized beings," settles the exact "territory" of the Wolffian shift in embryology.[55]

Now we understand how Kant departed from the doctrine of purpose that he held in the *Critique of Pure Reason*. Purposiveness in the *Critique of Judgment* is not just a convenient language, with which to describe the required systematicity of a nature founded on mechanistic grounds in order to articulate results of multiple determinative judgments. Rather, purposiveness is a specific concept emerging from the reflexive kind of judgment, under which a peculiar ontological kind is inscribed: living beings as organized beings. This conceptual shift accounts for the solution to the generation dilemma, stated theoretically in 1763, avoided in the First *Critique*'s Transcendantal Dialectics, and elaborated in the *Rassenschriften* on the grounds of an epistemological territory occupied by Wolffian embryology. The gap between the purpose-as-language of the First *Critique* and the teleological requisite of the *Rassenschriften* that Kant elaborated theoretically between 1775 and 1788 (*Über den Gebrauch der teleologische Prinzipien*) has been filled by this acknowledgment of embryological force and time as autonomous epistemological matter.

Of course, the idea of purposiveness seems closer to Blumenbach; and scholars have read it this way, because the *Bildungstrieb* aims at a goal, specifically, the species kind. But the epistemological territory of Blumenbach's conception is Wolffian, if we remember that the possibility of a scientific description of epigenesis, articulated by the specification of an appropriate force, was the result of Wolff's *Theorie der Generation*. For this reason, we related Kant's shift toward an analytics of biology in the Third *Critique* with the Wolffian embryological theories, rather than directly with Blumenbach. Briefly said, although Kant is inspired by Blumenbach's doctrine, his theory draws on the new epistemological territory (which yields this doctrine as well as its Wolffian alternatives) settled by Wolff.[56]

The *Critique of Judgment* asserted a theory of finality, resolving the hesitation and doubts of the first formulation in 1764. But the two previously central components of purpose, system and technique, are now distributed differently. The First Introduction spoke of "nature's technique" and about the teleological products that are "organisms."[57] Yet this technique needs no explicit assumption of an intention: because of the difference between artificial purpose and natural purpose, teleology refers to a merely regulative "idea of a whole" and implies no real intention as a cause. Finality, in the Third *Critique*, is divorced from the meaning "intentional," allowing Kant, later in the book, to reject any natural theology. "In order to keep physics strictly within its bounds, we there abstract entirely from the question as to whether natural purposes are purposes intentionally or unintentionally, since otherwise we could be meddling in extraneous affairs (namely, those of metaphysics)."[58] Because purposes in biology are purposes "of nature," the question of their intentional character is bracketed: "the very expression, purpose of nature, is sufficient to guard against this confusion. It keeps us from mingling natural science and the occasion it gives us to judge its objects teleologically, with our contemplation and then with a theological derivation."[59] Intention would be constitutive in the sense that we would assert: "this organism is there for such and such intention"; but this implies that the organism is something other than a product of nature, and we exit natural science.[60] In effect, nature as such can have no purpose.[61] The Critique of teleological judgment conceives of "purposiveness without intention" quite akin to the "purposiveness without purpose" by which the Critique of esthetical judgment defined Beauty. Hence this technique will not lead to God: it is no longer a craftsman scheme of the constitution of the world, like the "argument from design" criticized by Hume.[62]

Thus nature's technique is not particularly opposed to a "system." The First Introduction to the *Critique of Judgment* states that nature "logically specifies its universal laws into empirical laws according to a logical system."[63] That is the presupposition of the reflexive faculty of judgment. This is a new reading of the theory of a rational order of nature that is exposed in the Transcendental Dialectics. The system of nature is in itself "technical," and this technique is now opposed to mere "mechanical" development, namely, the determinative application of nature's laws, which would be exemplified in the doctrine of preformationism if it were true.[64] And those products of the technique of nature—the organisms—are organizing themselves. Consequently, there is a synthesis between the two initially opposite conceptions of purposiveness (system and technique), instead of the previous reduction of technique to order that Kant achieved in the First *Critique*. The organisms, as products of a *technique*, organize themselves *in a system*, because where the "idea of a whole" is concerned—in excess of the mere laws of nature—the organization spreads spontaneously in a kind of system, since no part was made for itself separately. Thus, the acknowledgment of the epistemological specificity of biology, emphasizing the meaning of epigenetic features in the definition of organism as self-organizing (even if Kant in his own biological theories was not as epigenetic as Herder or even Caspar Wolff),[65] led to a redefinition of the major concepts used in 1763 to express finality: system and technique.[66]

This shift is the reason for the new formulation of finality in 1791: what was known as the *concept of reason* in the Transcendental Dialectics appropriately becomes the *concept of faculty of judgment*. Indeed, finality in the First *Critique* was still seen as a craftsman scheme. That is why the systematic unity of nature, mechanically grounded, could be expressed in this language, namely, in a craft metaphor: "God created a harmonious universe, creatures, etc." But as soon as Kant distinguished this exterior, artificial purposiveness— the mere application of a pattern—from an immanent purposiveness, a self-organization, requiring the postulation of a "idea of a whole" to be known,[67] then such a technological purposiveness (proper to reason) ceased to be relevant, and we must substitute another concept for it: namely, purposiveness as immanent feature of the whole.

Now, it is well known that the reflexive judgment seeks the rule for the case; its paradigms are trials and medicine. Indeed, this judgment essentially presupposes a whole into which the cases order themselves. So finality as natural purpose, for example, organisms, must be grounded in this faculty rather than in reason.

Exterior purposiveness, such as technology—which is the territory of the finality as a concept of reason—appears in fact *twice external*: the pattern is in the constructor's head, *out of the object*; and the object often fulfills a task, therefore proving useful *for something else*. Of course, the paradigm of the new internal purposiveness, divorced from technology and hence from utility, lies in the *work of art*. In Kantian terms, *technique* no longer has the same relation to *system*, because now an *artistic*—rather than technological—scheme underlies this concept. The dual structure of the Third *Critique* is therefore not a matter of chance. Esthetics and biological teleology are the two faces of an un-technological, un-useful finality. But the biological side accounts for a new scientific territory delimited in the first place by the Wolffian embryology.

Georg-Ernest Stahl's *Theoria medica vera* (1712) contested the mechanistic account by inscribing within the organisms (and simultaneously coining this new word) a Reason unconsciously calculating its dissemination and internal moves; a little later, Albrecht von Haller provided to the living tissues the irreducible property of irritability, accounting for manifestations that no physical law of nature could by itself explain.[68] Almost in the same time, the French vitalists such as Bordeu[69] assumed a "proper life" to be inherent to the organs, named it "sensibility" and explained by this all-pervasive power the peculiar phenomena of living things such as pulse, secretions, and glands. Indeed, all those scientists were eventually contributing to a progressive substitution of the *external* idea of finality, instantiated *both* by the doctrine of a *soul* ruling the body and by a theological *economy of nature* through which animals and plants are designed for the service of others—by an *internal* purposiveness located inside the living body itself and having no analog in human technique. For this reason, if we remember this historical configuration, Kant's *Critique of Judgment* and its new concept of purposiveness could account for this immanentization of finality occurring at the same time in the life sciences.[70]

Conclusion: Rising Life Sciences and Metaphysics

From the beginning, the concept of purpose has been a key issue in metaphysics, specifically the order of nature. Since 1755 Kant carefully avoided any use of theological hypotheses in science. The most dramatic stumbling block was thus the problem of apparent order. The living realm, and primarily generation, was a crucial issue: therein, the contingency of order seemed irreducible and hence calling for theological support. This contingency seemed overcome in the *Critique of Pure Reason*, because Kant provided a methodological place for purposiveness, as a mere language, synonymous with mechanical lawlikeness and legitimized by the requisite of a universal nomothetic explanation of nature. But this was insufficient to address the claims of a newly arising science of life—a science to which Kant himself appeared to be a contributor through his writing on races in 1775–85. Within the transcendental framework, purposiveness had to be ontologized, in order for it to recognize the special status of biological phenomena. This

ontological weight of purposiveness means that the concern with the production of singular entities comes into play in the Third *Critique*, beyond
the systematicity of the laws of nature (sole issue addressed by the Appendix). This needed a shift through simply reflective judgment, as only able to
account for the specificity of biological, and mostly Wolffian, knowledge. By
the same movement, thus, all kinds of finality, included the formal finality
that is the pure systematicity of the laws of nature previously attributed to
pure reason by the Appendix, became part of the presupposition of the power
of reflective judgment. Evidence that this shift is effective between the First
and the Third Critiques is the following: "We must not consider unimport-
ant whether the expression 'purpose of nature' is interchanged with that of
divine purpose in the arrangement of nature, let alone whether the latter
is passed off as more appropriate and more fitting for a pious soul, on the
ground that surely in the end we cannot get around deriving the purposive
forms in nature from a wise author of the world. Rather, we must carefully
and modestly restrict ourselves to the expression that says no more than we
know—viz., 'purpose of nature.'"[71] This is exactly contrary to the text that I
quoted from the Appendix, assimilating Nature and God's wisdom. Because
now, purposiveness is no more a mere language about the general form of
nature (a language yielding some heuristics); it concerns our knowledge of
the production of some objects and cannot so easily be dealt with and attributed indifferently to various instances.

But if there is finality, this presupposes *contingency* vis-à-vis the mechanical laws of nature, since any process that is necessary according to these laws
constructs a whole from the parts, and therefore is not purposive. In the First
Introduction, Kant spoke of the "lawfulness of the contingent as such."[72]
By stating that prior to the mechanical action of forces there is an idea of
a whole, and epigenetic productive relationships between parts—although
those features are not really in the phenomena but must be enacted by
our judgment—the doctrine of the Third *Critique* gave life to this strange
locution, by showing how the contingent (regarding the universal laws of
nature), namely, the *form* of living beings, their conservation and transmission of forms in *self-repairing and heredity*, expressed a *kind of lawlikeness*,
even if this lawlikeness was not grounded in the physical laws of nature.
The embryological territory vindicated by Wolff, and his postulation of an
epigenesis that was scientifically accessible even though it was not in the
previous mode—Buffonian, etc.—with its proper descriptive methods and
its requirement of a force as an epistemological guarantee, allowed the construction of a concept of "natural purpose" that exposed this specific lawlikeness, namely: lawlikeness of the contingent forms regarding the laws of
nature. This lawlikeness can be conceived as a peculiar normativity, proper
to the living beings, because the difference between viable and nonviable

items is irrelevant with respect to the universal laws of nature (they have, in fact, the same kinds of causes); and therefore it appears contingent, and needs a transcendental account not yet available in the analytics of the laws of nature provided by the First *Critique*.[73]

Hence the metaphysical assumption of contingency, which did not refer to God because of its inherent lawlikeness, was achieved in the Third *Critique* through this investigation of the nascent life sciences. Kant was thus allowed to avoid a naturalism close to Spinoza's, for whom contingency would be a mere appearance, if not a mere language. The "lawlikeness of the contingent as such," expressed in the concept of "natural purpose" through the assumption of the main Wolffian epistemological features, finally enabled Kant to assess contingency without sacrificing to an all-disposing Creator the rational requirements, but with no commitments to a straight naturalism that, at this time, would have missed the specific epistemological and ontological character of the living beings.

These last remarks allow us to characterize more precisely Kant's contribution to biological thinking. In fact, to solve the metaphysical problem of the status of the contingent order, he mainly proposes a reformulation of the concept of purpose, which draws on (predominately) Wolffian epistemological configurations in embryology in order to give the living beings status as objects of science. Hence, he could fill the gap between metaphysical doctrine of purpose in the *Critique* and the use of purposive concepts in his essays on the natural history of races. Kant's shift between the First and the Third *Critique*, then, is intended to propose a synthetic concept of purposiveness and to take the philosophical measure of the Wolffian and Blumenbachian novelties in natural science. Thus Kant's theory of organisms as "natural purposes" is not a biological supplement to the *Critique*; rather, it is a conceptual solution to metaphysical problems raised by the many meanings of "purpose"—but a solution that, by taking into account some contemporary changes in the scientific theories and methods concerning life, contributed to establishing and delineating the metaphysical basis and epistemological territory of the rising biology.[74]

Notes

1.   Paul Guyer formulates an account of this shift from the First to the Third *Critique* through the First Introduction to the *Critique of Judgment*, rightly highlighting the focus on empirical laws of nature that characterizes the latter ("Reason and Reflective Judgement: Kant and the Significance of Systematicity," *Noûs* 24, no. 1 (1990): 36). However, I argue here that a complete account of this shift must consider the status of teleological judgment in the reelaboration of a concept of purposiveness. On this shift, see also Gérard Lebrun, *Kant et la fin de la métaphysique* (Paris: Colin, 1970), ch. 10, §§1–5.

2. Ak. 1, 227.
3. Ak. 1, 122.
4. Ak. 2, 106.
5. Ak. 2,107.
6. Ibid. This fact allows the living entities to be prima facie—like in the most orthodox tradition—a firm basis for the physico-theological proof: "in the building [*Bau*] of an animal, the organs of sensibility are so artistically joined to those of voluntary motion, and to the vital parts, that, unless one is of bad will [*boshaft*] (because no man could be so irrational), one is soon led to recognize a wise original being who brought about in such an extraordinary order those matter, of which the animal body is composed" (Ak. 2, 125).
7. Ak. 2, 117.
8. Ak. 2, 126.
9. "In any investigation of the causes of such and such effect, one has to cautiously preserve as much as possible the unity of nature, e.g. to bring back to an already known cause the various new effects" (Ak. 2, 113).
10. "We are able to have a concept of the way a being could cause some existence [*Wirklichkeit*], by analogy with what the men ordinarily make, but we cannot have a concept of the way a being could contain the ground of the internal possibility of the other things" (Ak. 2, 153).
11. Ak. 2, 126.
12. "It would be absurd to consider the first generation of a plant or an animal as a mechanical consequence of the laws of nature" (Ak. 2, 114).
13. On preformationism, see the classical study by Jacques Roger, *Les sciences de la vie dans la pensée française au 18ème siècle* (Paris: Colin, 1963). Peter J. Bowler in "Preformation and Pre-existence in the Seventeenth Century: A Brief Analysis," *Journal of the History of Biology* 4, no. 2 (1971): 221–44, provides an interesting account of the rise of preformationism, while Bodemer describes a state of the field that could illuminate Kant's own apprehension of preformationism ("Regeneration and the Decline of Preformationism in Eighteenth Century Embryology," *Bulletin of the History of Medicine* 38 (1962): 20–31). Michael Hoffheimer's analysis of the critique of preformationism in the framework of Maupertuis's thinking could be compared to Kant's criticism ("Maupertuis and the Eighteenth-Century Critique of Pre-existence," *Journal of the History of Biology* 15, no. 1 (1982): 119–44).
14. Ak. 2, 135.
15. Ak. 2, 114. On Buffon, see Phillip Sloan, "Organic Molecules Revisited," in *Buffon 88*, ed. Jean Gayon (Paris: Vrin, 1992), 415–38; Shirley Roe, "Buffon and Needham: Diverging Views on Life and Matter," in *Buffon 88*, 439–50. On Maupertuis, see the account by Mary Terrall, *The Man Who Flattened the Earth* (Chicago: University of Chicago Press, 2002).
Tonnelli paper's on the metaphysical problem of the unity of laws of nature at this period helps us to understand the relationship between Kant and Maupertuis, often quoted in the 1763 essay (Giorgio Tonnelli, "La nécessité des lois de la nature au 18ème siècle et chez Kant en 1762," *Revue d'histoire des sciences* 12 (1959): 225–41).
16. A654/B682. Let us notice that this order of nature is no mere subtlety to supplement the epistemological construction of the Analytics. Without order

in nature, the understanding could not compare anything—even with its principles, described in the Analogies, Postulates, etc.—and therefore it could not judge. "Considering [the empirical truth] we must presuppose the systematic unity of nature as objectively valid and necessary" (A654/B682). Notwithstanding their differences in interpretation, Kitcher ("Projecting the Order of Nature," in *Kant's Philosophy of Physical Science*, ed. Robert Butts (Dordrecht: Reidel, 1986), 201–35) and Gerd Buchdahl ("Causality, Causal Laws and Scientific Theory in the Philosophy of Kant," *British Journal for Philosophy of Science* 16 (1965): 186–208) do agree that the principles of the form of nature exposed in the Analogies are not sufficient to account for the possibility of a science of nature as such, and have to be supplemented by the principles of reason provided in the Appendix to the Transcendantal Dialectics. In the most recent survey of this debate, Grier argues in agreement with us in the end that since "Kant's general view is that the assumption that nature is systematically unified is always and already implicit in the theoretical undertakings of the understanding," hence "what will count as knowledge will ultimately be determined . . . by whether it accords with this transcendental assumption of pure reason" (Michelle Grier, *Kant's Doctrine of Transcendental Illusion* (Cambridge: Cambridge University Press, 2001), 282).

17.   My interpretation of the role of reason in the Dialectics, and the objectivity of the order of nature, which could seem strange to a reader accustomed to see the first part of the *Critique* as the main positive doctrine, is very sympathetic to the one developed by Kitcher, "Projecting the Order of Nature." It has also been inspired by Lebrun, *Kant et la fin de la métaphysique*.

18.   "If we survey the conditions of our understanding in their entire range, then we find that what reason quite uniquely prescribes and seeks to bring about concerning it is the *systematic* in cognition, i.e., its interconnection based on one principle. This unity of reason always presupposes an idea, namely that of the form of a whole of cognition which precedes the determinate cognition of the parts and contains the conditions for determining a priori the place of each part and its relation to the other" (A645/B673). The same concept of the "idea of the whole" will be reintroduced by Kant in the Third *Critique* in order to specify the patterns of our judgment about organized beings; on this, see below.

19.   System, purpose, design, architectonics, are words belonging to the same semantic network for Kant. The "Methodology of Pure Reason" will often make use of this last word, as is well known.

20.   Jonathan Bennett complains that Kant holds two separate concepts of God, one derived from teleology, the other resting on the idea of a unity of nature (*Kant's Dialectic* (Cambridge: Cambridge University Press, 1974), §86, 273). The link between them seems to him quite artificial; but this link is fully justified when we focus on the genesis of this idea of systematicity, stemming from the critique of abusive teleology in the *Beweisgrund* and requiring a more systematic conception of purposiveness.

21.   "This mistake can be avoided if we do not consider from the viewpoints of ends merely a few parts of nature, e.g. the distribution of dry lands, its structure and the constitution and situation of mountains, or even only the organization of the vegetable and animal kingdoms, but if we rather make the systematic unity of nature

entirely universal in relation to the idea of highest intelligence. For then we make a purposiveness in accordance with universal laws of nature the ground from which no particular arrangement is excepted, but arrangements are designated only in a way that is more or less discernible by us; then we have a regulative principle of the systematic unity of a teleological connection, which, however, we do not determine beforehand, but may only expect while pursuing the physical-mechanical connection according to universal laws" (A692/B720).

22. Kant seems here to refer to the general methodology of physiology since Fernel to the end of the eighteenth century, such as Andrew Cunningham reconstituted it ("The Pen and the Sword: Recovering the Disciplinary Identity of Physiology and Anatomy before 1800–II: Old Physiology—The Pen," *Studies in History and Philosophy of Biological and Biomedical Sciences* 34, no. 1 (2003): 51–76). The physiologist relies on a general assumption of design in nature, which allows him or her to deduce some functional mechanisms from the knowledge of the anatomist. Kant's concern to contain the general assumption of purposiveness in the boundaries of the needs of scientific inquiry, instead of drawing theological conclusions from it, clearly echoes the difference that should be made, according to Cunningham, between an assumption of design (teleological argument), from which theorical knowledge about particular features could be deduced, and a "natural theology," namely inference from the whole of nature to an all-powerful Creator. Physiology allows—and requires—only the former (and, historically speaking, it was mostly the natural historians rather than the physiologists who were concerned by this natural-theological inference).

23. *Critique of Pure Reason*, A699/B727. Translations are from the Cambridge edition of the complete works of Kant, *Critique of Pure Reason*, tr. Paul Guyer and Allen Wood (Cambridge: Cambridge University Press), 1998.

24. Therefore, I don't think that the antinomy of teleological judgment is somehow anticipated in the Third Antinomy, opposing natural causality and free causes. Further, as Peter McLaughlin indicated, the former involves speaking of causality on terms of wholes and parts, whereas the latter opposes two kinds of causation defined in terms of before and after (Peter McLaughlin, *Kant's Critique of Teleology in Biological Explanation: Antinomy and Teleology* (Lewiston: E. Mellen Press, 1990), 154). For a comparison of those two antinomies, see Renate Wahsner ("Mechanism—Technizism—Organism: Der epistemologische Status der Physik als Gegenstand von Kants *Kritik der Urteilskraft*," in *Naturphilosophie in Deutschen Idealismus*, ed. Gloy Karen and Burger Paul (Stuttgart: Fromann Holzboog, 1993), 1–23), who conflates the two, and John MacFarland (*Kant's Concept of Teleology* (Edinburg: Edinburg University Press, 1970)), who points out differences; for a general study of the antinomies in Kant's thinking, see Victoria Wike, *Kant's Antinomies of Reason, Their Origin and Their Resolution* (Washington: University Press of America, 1982); Henry Allison, *Kant's Transcendental Idealism* (New Haven: Yale University Press, 2003); Grier, *Kant's Doctrine*; and Brigitte Falkenburg, *Kants Kosmologie*.

25. *Critique of Pure Reason*, A688/B716.

26. *Vorlesungen über Rationaltheologie*, "Philosophische Religionslehre nach Politz," Ak. 28, 2, 2.

27. And this is also the case in the first part, as far as the "purposiveness without purpose" in the esthetics is also formal.

28. This last word was used in the *Beweisgrund*, as we have seen it in a previous quotation; it therefore connotes scientific considerations embedded in the genetical epistemology of this text—i.e., always looking at the systems that are about to naturally account for many effects—and hence seems less teleological than *Keime*. On this vocabulary, see the analysis by Zöller, "Kant on the Generation of Metaphysical Knowledge," in *Kant: Analysen—Probleme—Kritik*, ed. Hariolf Oberer and Gerhardt Seel (Wurtzburg: Königshausen and Nordmann, 1988), 71–90, and Sloan, "Preforming the Categories."

29. "What has to propagate must have been settled, in the reproduction force, as a determination previous to any occasional development adequate to the circumstances into which the creature may be involved, and into which it has to continuously maintain itself. . . . External things can be occasional causes, but not efficient causes, of what is necessary transmitted and propagated" ("Von den verschiedenen Rassen," Ak. 2, 435). One can notice that this conception of the milieu as occasional cause, which preserves the hereditary dispositions from any influences, opposes Buffon's theory of races as produced by degeneration of the germs by the milieu. It also opposes Blumenbach's *Handbuch der Naturgeschichte*, which followed Buffon on this matter. This is interesting to notice because in this little essay Kant begins by subscribing to Buffon's criterion of species (i.e., interfecondity).

30. On Kant's theory of heredity and generation, see Timothy Lenoir, *The Strategy of Life: Teleology and Mechanism in Nineteenth-Century German Biology* (Dordrecht: Reidel, 1982); Sloan, "Preforming the Categories"; Robert Richards, "Kant and Blumenbach on the *Bildungstrieb*: A Historical Misunderstanding," *Studies in History and Philosophy of Biology and biomedical Sciences* 31, no. 1 (2000): 11–32; Kenneth Caneva, "Teleology with Regrets," *Annals of Science* 47 (1990): 291–300; Clark Zumbach, *The Transcendant Science: Kant's Conception of Biological Methodology* (Den Haagen: Martinus Nijhoff, 1984); and other references in the Introduction to this volume.

31. Hence it will be no surprise to see the "technique of nature" as a main concept of the First Introduction to the *Critique of Judgment* (Ak. 20). More on this in Philippe Huneman, *Métaphysique et biologie: Kant et la constitution du concept d'organisme* (Paris: Puf, forthcoming).

32. "Über den Gebrauch teleologischer Prinzipien in der Philosophie," Ak. 8, 180.

33. *De formatione intestinorum*, §155. For Wolff's methodology in the framework of Christian Wolff's rationalism in Germany's late eighteenth century, see Shirley Roe, "Rationalism and Embryology: Caspar Friedrich Wolff's Theory of Epigenesis," *Journal of the History of Biology* 12, no. 1 (1979): 1–43. In *Matter, Life and Generation: Eighteenth-Century Embryology and the Haller-Wolff Debate* (Cambridge: Cambridge University Press 1980), Roe explains why the controversy that pitted Wolff against Haller had ideological and methodological dimensions too, so that neither author could really convince the other. See also Dupont (chapter 1 in this volume) on Wolff's final victory. Wolff wrote the *Theorie von der Generation* (1764) after Haller's recension of the *Theoria generationis* (1758); it is a translation of this book, but very much augmented with answers and refutations.

34. I am here describing something akin to what Nicholas Jardine in *Scenes of Inquiry: On the Reality of Questions in the Sciences* (Oxford: Clarendon Press, 1991)

called a shift in the scene of inquiry, even if he addressed the life sciences of the period just to come.

35. The role of the heart is very important for the Hallerian preformationism: it is for Haller the principle of life: "the animal could never have been without a heart" (*Prima linea physiologae* (Lausanne: Grasset, 1747; English translation, London, 1786), 585). Hence, even if we do not see it at the beginning, it must be there nonetheless, but with another shape, and invisible to us. Let us notice that Haller (and his friend Charles Bonnet, in *Considerations sur les corps organisés*) sustained a reformed preformationism, which means that the germs of an organ did not have to have the exact shape of the adult organism: they were principally a much more folded version of the organ.

36. *De formatione intestinorum*, §155.

37. See Stéphane Schmitt, *Les textes embryologiques de Christian Heinrich Pander* (Paris: Brepols, 2003), which contains a translation of Pander's *Beiträge zur Entwiclungsgeschichte des Hünhchens im Eye*.

38. See *Theorie von der Generation*, §§35 ff.

39. Ibid., §34.

40. On this point see Zammito, chapter 2 in this volume.

41. *Theorie von der Generation*, §13.

42. The model of hydraulics is pervasive in Lamarck's *Philosophie zoologique* (1809). By pointing out the epistemological differences between Newtonian methodology, which was dominant throughout eighteenth-century science, and Wolffian epistemology, my analysis follows the general diagnostic uncovered by Peter Hanns Reill, "Between Mechanism and Hermeticism: Nature and Science in the Late Enlightenment," in *Frühe Neuzeit-Frühe Moderne?*, ed. Rudolf Vierhaus (Göttingen: Vandenhoeck und Ruprecht, 1992), 393–421. This author shows that an "alternative to Newtonian mechanism" (p. 399) had been widely formulated in the mid-century, focusing mainly on life sciences and emphasizing a "historical" dimension in science, allowing scientists to treat vital phenomena and social matters. The historical understanding is salient in Wolff's new kind of science, as we see.

43. On Blumenbach and this parallelism, see Lenoir, *Strategy of Life*, and "Kant, Blumenbach and Vital Materialism in German Biology," *Isis* 71 (1980): 77–108.

44. Ak. 8, 159–84.

45. One can here refer to the careful study of those theories in Roe, *Matter, Life and Generation*.

46. *KU*, §81. Translations are from the Pluhar translation (*Critique of Pure Reason*, tr. Werner Pluhar (Indianapolis: Hackett, 1996)).

47. *KU*, §65, Ak. 5, 374.

48. Blumenbach objected to Wolff that his "force," *Kraft*, could not account for the conservation of form, e.g., for the fact that a pair of cows generate a cow, because *vis essentialis* is supposed to be the same in all species. Hence, Blumenbach postulated a *Bildungstrieb*, meaning by this *Trieb* the idea of a goal proper to the force: the *Bildungstrieb* in a given species is proper to this species, namely, it is a *Trieb* leading to the species' shape in the offspring. Wolff later elaborated his theory of *material qualificata* in response to Blumenbach, but their two conceptions had been published together with a paper by Born in 1781. Blumenbach exposed his theory in a text

entitled *Uber das Bildungstrieb*, which surely impressed Kant. On this controversy, see Lenoir, *Strategy of Life*; Robert J. Richards, *The Romantic Conception of Life* (Chicago: Chicago University Press, 2002); and Dupont (chapter 1 in this volume).

49.   *KU*, §65, emphasis added.

50.   Among the earlier attempts to capture the idea of an organism, Leibniz understood the difference between an organized being as a living being and an organized being as a machine in terms of degrees: an organism is infinitely organized, whereas a machine is finitely decomposable into parts that are not themselves organized. However, Kant rejected this interpretation in the Second Antinomy of the *Critique of Pure Reason*. See Grier, *Kant's Doctrine*, and Brigitte Falkenburg, *Kants Kosmologie. Die wissenscheftliche Revolution der Naturphilosophie I, 18.Jahrhundert* (Frankfurt: Klostermann, 2000). I analyzed this controversy in my "Kant's Critique of the Leibnizian Theory of Organisms: An Unnoticed Cornerstone for Criticism." *Yeditepe'de Felsefe (Istanbul hilsophical Review)* 4 (2005): 114–50.

51.   Zumbach, *Transcendant Science*, thus rightly emphasizes the epigenetic character of the Kantian definition of life.

52.   "Strictly speaking, therefore, the organization of nature has nothing analogous to any causality known to us" (§65, 375). So, what lacks an analogue is the way nature, i.e., some natural whole, organizes itself; but the production of the whole that self-organizes should be conceived by us analogously as our technique. This explains why Kant sometimes calls the causation of organized beings analogous ("a regulative concept for our reflexive judgment, allowing us to use a remote analogy with our own causality in terms of purpose generally," 175), while at other times he says they have no analogue.

53.   On the regulative/constitutive distinction, see the references in the Introduction to this volume, §1.4.

54.   The simplest characterization of a reflexive judgment is: find the rule for the case, whereas a determinative judgment would be: find the case for the rule. However, the main point is that there is no rule given prior to the act of judgment in the case of a simply reflexive judgment. This means that the judgment itself has to state the conditions under which it has to be enunciated—that is why Kant calls it "reflexive." This implies that there is a deep affinity between what is called "critique"—as an investigation on the conditions of knowledge—and "reflexive judgments," an affinity that is considered in the First Introduction to the *Critique of Judgment*. Kant says, then, that teleology cannot be a dogmatic discourse, but it has to be part of a critical knowledge (*KU*, Introduction, Ak. 5, 194).

On "simply reflexive judgment" distinguished from "reflexive judgments," see Béatrice Longuenesse, *Kant et le pouvoir de juger* (Paris: Puf, 1993), 202ff.

55.   In a paper that I read after having elaborated this one, Joan Steigerwald establishes very similar conclusions, namely, the link between Kant's emphasis on reflective judgment as the domain of purposiveness and Wolff's changes in the structure and aims of embryology ("Instruments of Judgement: Inscribing Processes in Late 18th-Century Germany," *Studies in History and Philosophy of Biology and Biomedical Sciences* 33 (2002): 79–131). Steigerwald's argument mostly relies on the novel characteristics of the experimental activities and their specific relation to embryogenesis as their object. "Kant argued that in organized bodies, we are confronted with natural

phenomena to which we are not able unconsciously and automatically to apply a priori concepts, but rather are forced to reflect upon which concepts are appropriate. Investigators such as Blumenbach, through their experiments and through notions such as the *Bildungstrieb*, provided instruments assisting these reflective judgments, giving us means for thinking concretely about how such natural products arise, imagining the generative process by inscribing it in organic matter, without claiming that these reflections are actually determinative of those products" (p.119).

56. The connection between Kant and a specific territory of embryology, defined by specific experimental procedures and explanatory requirements (and shared by Wolff and Blumenbach), rather than Blumenbach alone, is also clearly assumed in Steigerwald, "Instruments of Judgments," 98.

57. Ak. 20, 217—the "internal structure of plants and animals" is due to nature's technique.

58. *KU*, §68, 382.

59. *KU*, 382. Here, as in the Appendix, Kant echoes the separation that Cunningham detected in the structure of seventeenth- and eighteenth-century physiology, namely, the difference between a teleological argument supporting scientific deductions and a natural-theological inference.

60. More precisely, we exit any kind of knowledge of nature, *Naturwissenschaft* as well as *Naturlehre*, two terms that are somehow embraced by *Naturforschung*.

61. Kant's strategy here is rather delicate. Briefly said: because we know that nature can have no purposes, we are allowed to speak of the intentions of nature, since it is meaningful. "That is why, when in teleology we speak of nature as if purposiveness in it were intentional, we do so in such a way that we attribute this intention to nature, i.e., to matter. This serves to indicate that this term refers here only to a principle of reflective, rather than of determinative, judgment. (It indicates this inasmuch as no one would attribute to a lifeless being an intention in the proper sense of the term, and so no misunderstanding can arise" (Ak. 5, 383).

62. Of course, Kant authorizes the use of technological expression—but it is always by analogy, and this analogy is, we might say, "neutralized," because Kant had previously demonstrated that the expression "the intentions of the matter" is, rigorously speaking, absurd. It doesn't mean anything in the natural sciences, hence it only indicates a kind of epistemological requisite. "I agree that we speak, in the teleology of nature, as if purposiveness were intentional in it, but in order to ascribe this intention to nature too, and therefore to the matter, one will need to show (since about this *no misunderstanding is possible, nobody being able to ascribe to inanimate matter an intention in the proper sense*) that this word refers only to a principle of the reflexive faculty of judgment, and not a principle of the determinative faculty of judgment, and that it must not introduce any particular principle of causality" (*KU*, §68, Ak. 5, 383, emphasis added).

63. "First Introduction," Ak. 20, 216.

64. Ak. 20, 217.

65. Zammito, chapter 2 in this volume, and Frederick Beiser, *The Fate of Reason* (Chicago: University of Chicago Press, 1990), 150–55.

66. This conceptual shift illuminates partially Kant's puzzling "epigeneticism." When those two concepts (system/technique) are barely opposed, as in the *Beweisgrund*,

then epigenesis (immanent organization by the system of laws) and preformationism (initial technical creation by God) are in crude opposition. But when the conceptual work of the Third *Critique* proposes a kind of synthesis in this new concept of finality—free from the craftsman scheme—then this crude opposition is no longer valid. That's why Kant, in this framework, named his own position on generation "generic preformationism," which means an epigenesis presupposing an "original organization" undeducible by natural laws from mere matter (*KU*, §81). See Mark Fisher, chapter 4 in this volume, for more on this point.

67. The distinction between "relative" (i.e., external) and "internal" finality is made in *KU*, §63. Here, Kant shows that there the former, based on considerations of utility, is not theoretically legitimate. Examples of it are of course the traditional material of natural theology: plants intended to nourish animals, etc.

68. Haller, *Elementa physiologae* (Lausanne, 1757).

69. Bordeu, *Recherches sur les glandes* (Montpellier, 1755).

70. This analysis takes account of the progressive recognizing of an internal power proper to living beings: Jacques Roger, *Les sciences de la vie*, reconstituted the intellectual landscape of this recognition, a landscape with which Kant was quite familiar. Bodemer investigates along those lines the role of Trembley's experiments on the regenerating polyp (around 1732), accounting therefore for the rise of epigeneticism in the second part of the century. Reill analyzes this intellectual movement as the Enlightenment response to the skepticism toward mechanicisim and Newtonian methodology ("Unlike the neo-mechanicists, they sought to reformulate the concept of matter, along with those of force, power and connection in their construction of a science that respected natural variety, dynamic change and the epistemological consequences of scepticism" ("Vitalizing Nature and Naturalizing the Humanities in the Late Eighteenth Century," *Studies in Eighteenth Century Culture* 28, ed. J. C. Hayes and T. Erwin (Baltimore: John Hopkins University Press, 1999), 365)).

71. *KU*, §68, 382.

72. Ak. 20, 217.

73. Such an interpretation has also been recently worked through by Hannah Ginsborg, "Kant on Understanding Organisms as Natural Purposes," in *Kant and the Sciences*, ed. Eric Watkins (Oxford: Oxford University Press, 2001), 231–59.

74. This chapter was written when I had access to Peter Hans Reill's last paper on the subject (Reill, "Between Preformation and Epigenesis: Kant, Physicotheology and Natural History," in *New Essays on the Precritical Kant*, ed. Tom Rockwell (New York: Humanity Books, 2001), 161–81). I have to acknowledge here the fact that we came to quite convergent results regarding the place of epigenesis in Kant's thought, and his way out of the 1763 generation dilemma.

4

Kant's Explanatory Natural History

Generation and Classification of Organisms in Kant's Natural Philosophy

Mark Fisher

Abstract

There is a long tradition of Kant interpretation that takes seriously the importance of Kant's engagement with the sciences for understanding the aims and methods of his philosophical project. Given Kant's own explicit statements about the sciences, it is no surprise that much of the literature in this tradition focuses primarily on Kant's relationship to mathematics and physics. That is, the a priori element required by Kant for a discipline to deserve the honorific title of a science is most obviously met by mathematics (which is wholly a priori), and then, only slightly less obviously, by physics (which has both an a priori and an empirical element). What is somewhat surprising, however, is the relative lack of attention to Kant's engagement with another discipline that seems to concern him at least as much throughout his most productive philosophical years as does physics, namely, natural history. In this chapter, I not only suggest that natural history plays a more significant role within Kant's philosophy than is generally acknowledged, but I also argue that Kant's proposal for transforming natural history from a primarily descriptive discipline into an explanatory science represents a plausible answer to some of the most troubling questions, especially concerning generation and classification, that face practicing naturalists at the end of the eighteenth century.

Introduction

In §81 of the "Critique of Teleological Judgment" Kant claims that although we cannot understand the generation of organized beings according solely to what he refers to as the "mechanism of nature," we must not exempt these beings entirely from this mechanism if we are to continue viewing them as products of nature.[*] In the process of explaining why this is the case, he provides what may appear to be simply a summary and evaluation of the various alternatives offered in the eighteenth century for explaining how organisms are generated. He takes the theories of preformation and epigenesis to be the only legitimate possibilities for providing this explanation, and then voices support for the theory of epigenesis on the basis of both experience and reason. In so doing, however, Kant makes a claim that should strike anyone familiar with the various debates concerning generation in the eighteenth century as odd, if not simply mistaken. That is, he claims that the theory of epigenesis, which for much of the preceding two centuries was considered a rival to the theory of preformation, is really a version of this theory.

In what follows, I suggest that we read this claim of Kant's as part of an original, defensible position within the eighteenth-century debate concerning the generation of plants and animals. According to this position, the best way to account for the generation of *individuals* in the organic realm is according to the theory of epigenesis, that is, by claiming that the organic structure of the being emerges gradually from a previously undelineated mass through a process that is not caused or guided by external forces.[1] Representing the possibility that certain natural beings should have the ability actually to generate other natural beings, resembling them in significant ways but differing from them in each case, involves, however, seeing the *genus* of beings to which the individual belongs as having been preformed, that is, brought about precisely by such external forces.[2] The picture Kant has in mind is, roughly, as follows.

There is a contingent connection of various parts (i.e., organs) serving their characteristic functions within an organism that is not the same kind of connection that characterizes the part-whole relation in material bodies more generally. Accordingly, we are not able to explain the possibility of the generation of an individual organism by reference merely to the general laws of motion that constitute an essential element of our understanding of matter *as such*. Nor are we able to understand particular processes within individual organisms as governed solely by laws that can be understood as more particular instances of these general laws. Matter, motion, and the fundamental forces of attraction and repulsion that give rise to these are alone sufficient for understanding neither the ultimate historical origin of organic beings nor the actual functioning of empirically given individual organisms.[3]

[*] I would like to thank all the members of the panel at the Eleventh Quadrennial Congress of the International Society for Eighteenth-Century Studies, in August 2003 at UCLA, for the comments and discussion of the paper given there from which this chapter stems, and especially Philippe Huneman for inviting me to participate in the panel.

On this view, a certain kind of epigenetic explanation is deemed impossible, namely, the kind that attempts to explain the ultimate origin of organized bodies by beginning with the mechanically explicable "tendency" toward order of matter *as such* and claiming that, over time, this "tendency" leads to the kinds of complex orders that are observed in the organic realm.[4] According to Kant, such speculative attempts to turn reason's *idea* of a "great chain of beings" into a physical natural history are supported neither by reason nor by experience.[5]

If we are to understand the generation of beings as complex, individual organisms according to natural laws (rather than as products of art or of an external causality), we have to assume a contingent, original organization from which these laws follow. That is, the best way of understanding the possibility that organized beings are generated naturally involves the claim that "some individual members of the plant and animal kingdoms, although immediately formed by God and thus of divine origin, possess the capacity, which we cannot understand, actually to generate their kind in accordance with a regular law of nature."[6] Because the first members of each kind of plant and animal are assumed to be the result of divine artifice, *preformation* is a condition for the possibility that organisms be generated naturally. What the preformation of the *first* individuals allows, however, is a natural causal account of the generation of *subsequent* individual organisms, according to which both parent organisms contribute to the determination of which characteristics the developed organism will eventually exhibit. In other words, the preformation of the *genus* in the original, created pair is a condition for the epigenetic capacities of *individual* members of that genus.

I provide a sketch of this view in the following by beginning, in section 1, with a consideration of one of Kant's most enduring interests with respect to the natural world, namely, natural history. Attention to his views on transforming natural history from a primarily descriptive cataloguing of natural variety into a discipline seeking to provide physical, causal explanations of this variety shows that, from his earliest writings, Kant favors an epigenetic account of order in nature more generally. In section 2, I discuss the challenges to this general view that led Kant to adopt a preformationist view of the ultimate origins of plants and animals. Some consideration is also given to the problems with other versions of preformation that led Kant to adopt the particular kind of preformationist position that he did. I hope to show that Kant's position concerning the generation of organisms is of interest both philosophically and historically insofar as it brings together two important issues in the eighteenth-century investigation of the organic realm, namely, generation and classification, and addresses them from the standpoint of a suggestion for establishing an explanatory natural history.

1. Natural History in General

An enduring philosophical and natural scientific interest of Kant's is the attempt to understand and explain various aspects of the present state of the natural world as results of natural processes, the beginnings of which processes are projected backward indefinitely into the past. One of his earliest published works, the *Universal Natural History and Theory of the Heavens* (*Allgemeine Naturgeschichte und Theorie des Himmels*, 1755),[7] is concerned with providing just such an explanation of the most general features of the solar system, that is, an explanation of the shapes and masses of the planets, the trajectories they trace around the sun, and so on, as the results of the interplay of natural forces essential to matter over the course of time. Kant proceeds here according to the somewhat simple hypothesis that if the generally accepted laws governing the motion of material bodies are sufficient for accounting for the maintenance of a system as apparently complex as that of the planets and fixed stars, they may well be able to account for the generation of such a system as well.[8] Accordingly, arrangements in nature that end up serving the purposes of both humans and other natural beings need not be understood exclusively as the result of processes intentionally directed at these arrangements as ends. Many of these arrangements can be explained as beneficial, but necessary and unintended, consequences of the most general laws governing nature.[9]

This approach is aimed at reducing *our* reliance on intelligent design in providing an account of various complex phenomena in nature, while, at the same time, stressing the ultimate reliance of all things whatsoever on a being possessing infinite wisdom.[10] It accomplishes the former by rejecting the claim that natural laws in general, and the laws of motion in particular, can be trusted only to maintain an order divinely instituted in the cosmos. This claim is replaced by the idea that the generation and maintenance of order are not so different as to require radically different explanations. Accordingly, we are free to proceed in our natural scientific investigations *as if* nature alone were sufficient to bring about and maintain the immense order and harmony observed in the natural world. From a standpoint other than that of the natural sciences, however, we recognize that the very possibility that matter should order itself into such regular and harmonious arrangements points to an ultimate, intelligent ground of the entirety of nature.[11] To put Kant's point somewhat paradoxically, the less frequently we are required to call on special provisions in explaining the phenomena of nature, the more convinced we will be that the whole of nature constitutes a single system brought about according to a wise plan. Moreover, investigating nature according to the idea of a single, harmonious system, governed by natural laws and not requiring the frequent addition of supernatural influences, introduces an order into our knowledge of nature that would never come about were we to content ourselves with merely observing, describing, and recording facts about the great diversity found in the natural world.[12]

This description of Kant's methodology for investigating the generation and maintenance of natural order is taken from his pre-Critical works; however, regardless of what other aspects of his pre-Critical thought undergo revision as part of the "Critical turn," Kant's general understanding of the goals of natural history remain largely unaffected through the "silent decade" and into the Critical period.[13] If one looks at the various occasional essays Kant composed between 1775 and 1788 on topics central to eighteenth-century natural science,[14] as well as the notes taken by students from his lecture courses on physical geography[15] during this same period, it becomes clear that Kant remained concerned not only to develop further the methodology outlined above and to place it in contrast to other ways of investigating nature prevalent at the time, but also to apply this way of thinking about nature to some of the most difficult issues facing practicing naturalists at the end of the eighteenth century.

The following general picture of natural history and its role in human cognition emerges from these essays: Natural history (*Naturgeschichte*) shares with both "system of nature" (*Systema naturae*, *System der Natur*) and natural description (*Naturbeschreibung*) the concern to provide a systematic view of the diversity observed in nature. That is, each of these aims to discover and apply principles by which the seemingly endless variety of individual objects in nature, revealed both by personal experience and by the experience of others, can be ordered into a whole, the basic outlines of which can be grasped by our finite intellects. The primary difference between natural history and natural description is that the former takes as its primary object courses of events in the world (*Lauf der Welt*, *Weltlauf*), whereas the latter deals primarily with the showplace of the world (*Schauplatz der Welt*).[16] In other words, natural history deals with change over time, and natural description classifies the things in nature and their relations to each other at a given time.[17]

Natural history and natural description are distinguished from systems of nature through their shared concern with understanding nature as a physical, rather than a merely logical, system. That is, the various systems of nature, of which Kant mentions Linneaus' *Systema naturae* as an example, judge two or more things to be related to each other just in case each exhibits a common property or set of properties.[18] Relations between objects in the natural world, according to this view, are logical relations of similarity and difference (in *phenomenal* characteristics such as size, color, number, shape, and arrangement of observable parts) that are captured by conceptual classification according to genus and specific difference.[19] Understanding nature as a physical system, in contrast, involves judging two or more things to be related either (1) geographically, according to spatial relations among natural objects existing at the same time, or (2) historically, according to the places these objects occupy in a particular causal chain.[20] It is ultimately the combination of these two kinds of physical relation (i.e., the spatial and the temporal or the geographical and the historical) into a system that is involved in a true natural history, according to Kant. Geographical classifications and descriptions of the current state of the world are the *explananda* for which we seek a historical *explanans*. Accordingly, there is a two-fold relation between these attempts at system. On the one hand, the present state of the world as it is described in geographical terms by natural description is the result of the historical processes in nature that are the objects of natural history, and, on the other, the framing of hypotheses concerning which processes need to be assumed in order to account for the current state of the world, which is central to natural history, relies on the previous acquaintance with this state that is the goal of natural description. Another way of putting this is to say that natural description is the *ratio cognoscendi* of a physical order in nature for which natural history seeks the *ratio essendi*.[21]

Requiring as it would, first, a complete description of the character and relations of all natural objects at the present time and, then, an account of how these objects came to be as they are, a completed natural history can be only an idea that we endeavor to approximate as closely as we can given the incompleteness of our knowledge of nature. As it stands, however, not a great deal of progress is being made in this direction due to the prevalence of works in natural description that are being passed off as natural histories, the denial of a significant distinction between natural description and natural history, and the prevalence of logical, rather than physical, systems of nature.[22]

2. Generation and Classification

With this overview of Kant's thinking about the aims of natural history in place, we can now turn to consider the issue with which I began this chapter, namely, the issue of how Kant thinks we are to understand the processes by which organisms are generated. As is fairly well documented, thought on this issue in the seventeenth and eighteenth centuries is generally divided into two camps, that is, there are supporters of the theory of preformation, on the one hand, and supporters of the theory of epigenesis, on the other. The theory of preformation in its original form involves the claim that all individual organisms are created simultaneously by God, with future generations literally encased in each other and awaiting their proper time to begin unfolding.[23] What appears to be the production of a new, individual, organized body is really only the increase in size of an already existing organism. According to this version of the theory, each individual organism is of supernatural origin, but there is room for natural explanations concerning such things as the increase in size of the embryo and the further functioning of the preformed organic machine. The theory of epigenesis is contrasted with this theory by means of the claim that the generation of individual organisms occurs naturally through successive stages, whereby a relatively undifferentiated mass develops into an organized structure. What appears to be the production of a new individual organized body is actually just that.

At the time Kant turns his attentions to these matters, the general attitude within the German-speaking world is that some form of preformation must be involved in accounting for precisely those characteristics of plants and animals in virtue of which they are generally distinguished from other natural products; that is, their consisting not merely of parts but of organs. At least since the middle of the seventeenth century, it is generally accepted that all plants and animals originate from, or exist originally as, seeds or eggs.[24] Beginning with these generalizations from extensive observation—namely, that plants and animals are organized beings and that they originate from seeds or eggs—it is fairly easy to understand why many opted for the theory of preformation in explaining the phenomena associated with the formation and development of a new, or apparently new, individual. As Kant himself claims:

> How a tree, for example, should be able through an inner mechanical constitution to form and model the sap in such a way that, in the bud of the leaf or in the seed, something should come about that contains a similar tree in miniature, or out of which such a thing could come to be, can in no way be understood according to all of our knowledge.[25]

In other words, it is difficult (if not impossible) to imagine how an organized being can produce something as apparently undifferentiated as the seed or the egg is, that is nevertheless able to develop, through natural processes unguided by any plan, into an organized structure capable of maintaining itself as an individual, repairing itself in the event of injury, and, to the extent that the new individual shares the capacity of its progenitor, propagating the species.

For the purposes of a mechanical investigation of the processes by which individuals develop and maintain the structure they exhibit, most naturalists assume that it is necessary to presuppose that the seed is really not so simple to begin with, that is, that there is actually an articulated structure present in the seed, though unobservable, the growth of which through the appropriation of nutritive elements results in the observable structure we eventually perceive. The acceptance of such a view commits one to some form of *preformation*, but it alone does not dictate the acceptance of one form of this theory over another. That is, by the middle of the eighteenth century it is accepted that not all organs become visible simultaneously. Certain organs become visible well before others, and these latter appear to be generated successively from the former organs. Accordingly, the version of *preformation* that holds each individual to be fully articulated in miniature and simply to increase in size, that is, *individual preformation*, seems untenable. In light of these observations, a new version of *preformation* is suggested, according to which it is not the case that the seed or egg contains a fully formed individual in miniature. It contains, rather, preformed germs, or primordia of the eventual organs that are characteristic not of a completely determinate individual but of the kind or species from which the individual stems. This version of preformation, accepted with varying degrees of commitment by Haller and Bonnet, seems to have some advantages over the version that holds each individual to exist fully formed since the creation.[26] Not only is it able to incorporate observations that seem to support an epigenetic account of generation (i.e., the gradual appearance of structures in what previously appeared an undifferentiated mass), but it can also, at least to some degree, take what we would now call environmental considerations into account in explaining the particular ways in which individuals develop. According to this view, the individual is not completely determined by internal factors to unfold in the way it eventually does regardless of the contingent circumstances in which this unfolding takes place. Rather, features that have long been understood to be correlated with differences in the development of individuals of the same species—for example, the health of the mother and the climate in which the individual develops—can be incorporated as causal factors into our accounts of diversity in the organic world. Despite these advantages with respect to the theory of individual preformation, however, there remains a significant difficulty with this view; namely, it has difficulty accounting for the undeniable facts surrounding heredity.[27] Because the seed or egg is already present in the female prior to fertilization, no real determinative role is accorded the male with respect to the actual generation of the new individual; rather, the male provides only a catalyst for beginning the unfolding process.[28]

The theory of epigenesis, in contrast, leaves open the possibility that both parent organisms contribute to the original makeup of their offspring. Accordingly, this theory has an advantage over then-current versions of the theory of preformation. This advantage, however, should not blind us to problems that then-current versions of *epigenesis* face.

The two leading proponents of epigenetic accounts of generation in the middle of the eighteenth century are Buffon and Maupertuis.[29] Each of these thinkers attempts to account for the actual production of a new individual from previously existing individuals by outlining a process in which organic material is, first, taken from the various parts of each parent and mixed together in their respective reproductive materials. From the combination of these reproductive materials an embryo is formed that will eventually assume the form characteristic of the species, but likewise exhibit traits of both parents. The possibility that this should happen results either, as in the case of Buffon's account, from each organic molecule having been impressed with the form proper to the molecules making up the part of the organism from which it came or, according to Maupertuis, from each molecule returning to its proper place relative to the other molecules through capacities similar to memory, attraction, and repulsion. The chief advantage that such accounts of generation have over the theory of preformation is that they are far better equipped to deal with the growing body of undeniable facts concerning heredity that are being discovered in the eighteenth century.[30] The primary disadvantage is that it is not obvious that the *particular* claims concerning the *precise* mechanisms by which these facts are accounted for are true, coherent, or ultimately capable of accounting for the processes involved in the formation of a new individual.

These facts alone, however, do not require that we abandon the theory altogether and decide between individual preformation and the version of germ preformation put forth by Haller and Bonnet. Recognizing, correctly it seems on the basis of the terms in which this debate is couched in the eighteenth century, that none of the arguments adduced for the truth of one theory or for the falsity of the others is conclusive, Kant opts for a qualified endorsement of that theory that is most in line with his own metaphysical and methodological position concerning natural history, namely, the theory of epigenesis. The qualification of this endorsement, however, is such that it may appear that he is granting to the preformationist all that he or she requires, that is, that regardless of how far we are able to carry out our investigations of organic phenomena according to natural laws governing the interaction of material parts, eventually we will have to have recourse to a nonmaterial principle of design in accounting for the unity of these parts in an organic whole.

Kant's suggestion, which he will repeat in discussing the various possibilities for accounting for generation in the third *Critique*, is that any defensible epigenetic account of the generation of organized beings will differentiate itself from preformationist theories not by rejecting all appeal to external principles of design, but, rather, by getting clear on precisely where in our explanations these principles are required. His suggestion for grounding an epigenetic account of the generation of individuals is that, perhaps, "certain individuals of the plant and animal kingdoms, while themselves of supernatural origin, nonetheless have a capacity we cannot understand to generate their kind, and not simply to unfold them, according to a regular law of nature"; accordingly, we should "allow to the first divine arrangement of plants and animals a capacity truly to generate their like thereafter, and not merely to unfold them."[31] What this suggestion shares with other theories of preformation is commitment to the claims that (1) the first individuals of any particular species or genus are created directly by God, and (2) all further members of this species or genus develop according to natural laws from this initial arrangement. The primary difference between Kant's theory and others concerns the way in which claim (2) is interpreted. It seems that it can be interpreted in three ways, each of which gives rise to a different theory of generation.

If the claim is interpreted to mean that the further members are actually present, as fully formed miniatures, in the initial arrangement, and the laws of nature govern merely the manner of their gradual unfolding, then all individuals are created directly by God and one holds the theory of *individual preformation*. If, alternatively, it is interpreted to claim that the germs of future generations are initially present in or are generated naturally by one parent or the other prior to any intercourse, then one holds a *germ* theory of *generic preformation*. If, finally, we interpret this claim to mean that the first individuals are created with a capacity to produce a relatively undifferentiated mass through their intercourse, which then develops through stages into a new individual of the same species as its parents, then one holds a theory of *epigenesis* that can also be understood as a theory of generic preformation.

I think reading Kant as suggesting this third alternative serves not only to make good sense of his treatment of these issues in general, but also to make sense of his otherwise somewhat odd claim in the third *Critique*, that the theory of epigenesis, which had for so long been put forth in stark opposition to the theory of preformation, is really a version of this theory. Even if this is the right way to understand these features of Kant's thinking about the generation of organized beings, however, it is still not sufficiently clear precisely why Kant opts for the specific version of *generic preformation* that he does. Ultimately, the reasons for this are several, and I cannot treat each of them here, but I will draw attention to one that I think is particularly important.

As mentioned above, one of the primary virtues of epigenetic accounts of the generation of individuals is that, at least in the cases of what we now call sexual reproduction, both parent organisms contribute materials to the physical constitution of the offspring. Accordingly, the well-established fact that offspring tend, though not always in straightforward ways, to resemble their parents can be accounted for by such an account more satisfactorily than it can by its alternative. Further, the prima facie causal role played by factors external to the individual in accounting for variation between individuals of the same species need not be explained as a merely contingent, harmonious correlation. Not only are these facts better accounted for in this way, but the account provided also enables us to fix a physical, to some degree experimentally useful, criterion by which to make judgments concerning species. In other words, the answer to the above question concerning *generation* has important implications for another of the most contested issues concerning the organic world in the eighteenth century, namely, the proper basis for the *classification* of natural products.

Given Kant's interest in natural history, that is, in the systematic investigation of nature according to the physical, historical processes by which it has come to exhibit the diversity we currently observe in it, it should be no surprise that Kant would favor an account of the generation of organized beings that allows us to explain the features on the basis of which we classify them as resulting from *historical* processes. More specifically, Kant wants to establish connections among the logical system of nature, which subsumes individual natural products under class concepts according to outward similarities; the geographical system of nature, which reveals variations among members of classes that are correlated with the places on the earth in which they are found; and the historical system of nature, which seeks to explain both the unity and the variety observed among the individuals comprising these classes in terms of initial members of these classes and physical laws by which they produce further members that develop differently due both to internal, or hereditary, and external, or environmental, factors.

Kant uses the language of preformed germs and dispositions in his essays concerning issues of classification (and here I have in mind the 1775 essay "On the Different Races of Humankind" and the 1785 essay "Determination of the Concept of a Race of Humans") as part of a more detailed explication of how his model of generation, according to which some plants and animals, while themselves products of supernatural causality, might be able genuinely to produce new members of their kind, could be thought to function. This leads, understandably, to the idea that Kant's account is essentially the same as the germ preformation of Haller and Bonnet. I think before we agree with this assessment, however, we should consider the following. In these works, Kant is interested in making use of the notion of a physical species, at least in part, because it provides a criterion for membership in a species that is, at the same time, explanatory of the ability of its members to combine and produce fertile offspring. A species, according to this view, is not an aggregate of individuals contingently connected by means of their sharing certain merely phenomenal characteristics. Rather, for Kant, a species is a real whole, the parts of which are connected through a generative force that is (at least partially) explanatory of the unity of the whole, of the diversity of the individuals comprising the whole, and of the ability of these individuals truly to generate new members of the species. All of these features result from membership in a historical line of descent in which individuals develop differently according both to the differences in their natures and to the differences in the environments in which they develop. The differences in natures are, in turn, explained by appeal to the characteristics of their progenitors, and the differences in their environments are explained by appeal to the historical development of both other organic and nonorganic natural products.

According to this somewhat cursory suggestion for understanding Kant's position, he is concerned with providing a natural history that is able to account for the taxonomic distinctions (of genus, species, variety, etc.) that we form on the basis of empirically observed commonalities between individuals, and correlations between individuals and their environments, in terms of natural laws arising out of as few initial divine arrangements as possible—rather than as resulting from precisely as many divine arrangements as there are individuals we are concerned to classify. To establish this, individual organisms must be viewed as generated in time through natural causal processes involving (at least in cases of sexual reproduction) both parent organisms. The possibility that the parent organisms should be able to generate a new organism in this way stems from their being members of the same species, or having stemmed from the same divinely organized original pair. It is for these reasons that Kant holds the claims I attribute to him above, namely, (1) the best way to account for the generation of individuals in the organic realm is according to the theory of epigenesis and (2) attributing epigenetic generation to these individuals requires that we view the genus as being preformed.

Conclusion

In the above, I have attempted to trace the general outlines of Kant's position on the generation of organisms as I see him developing it from the pre-Critical writings through the third *Critique*. I have concentrated only on those aspects that appear to me to remain constant. This should not be taken to indicate that I don't think any of the developments in the natural sciences and in natural history that take place during the period in which Kant is writing have any effect on Kant's thinking about organisms, but only that I am committed to the project of understanding these changes in Kant's thinking against the background of what I take to be these fundamental commitments.[32] There is certainly a good deal more that needs to be said about Kant's grounds for rejecting both the individualist and the Haller-Bonnet germ versions of the theory of preformation, as well as those versions of the theory of epigenesis that claim to do away with all appeal to principles of intelligent design. Especially given the contemporary tendency to see the latter as the most promising direction for scientific inquiry into the reproductive capacity of organisms, Kant's insistence that such an account cannot work may seem unfortunate. I think, however, that closer attention to the versions that present themselves as options in the late eighteenth century would reveal that they promise something surely attractive for the naturalistically oriented researcher into the complexities of the organic world but that they actually deliver far less than they promise. Kant's own account, it seems, offers just as much as they do for our understanding of the natural, physical processes that characterize organisms, but without claiming that all of these without exception can be explained by reference merely to natural forces discovered in matter as such. Of course, sorting out the various ontological and epistemological issues at stake in this disagreement cannot be done here, but I hope to have at least set the stage for a more penetrating treatment of these issues.[33]

Notes

1. Harvey is the first to use the term *epigenesis* to describe this process. He does not pretend to be able to explain the possibility of this process, however, and restricts himself to a merely descriptive task. Later thinkers, such as Buffon and Maupertuis, try to provide explanations of this process that, as we see below, Kant cannot accept. This does not mean that he disputes the claim that organisms have the capacity to generate other organisms naturally or that this generation occurs as the theory of epigenesis describes it. Kant merely disputes that these thinkers have explained sufficiently how such a process is possible.

2. I claim that it is the genera that are preformed for Kant, rather than the species, for the reason that Kant seems open to the Linnean account, according to which the existence of various species can be accounted for by presupposing original members of the genera plus reproductive and geographical differences. This commitment is apparent in the essays dealing with race and teleology, and comes out as well in the Introduction to the *Critique of Judgment*. There, Kant talks about the supposition that underlies our classificatory endeavors being that nature specifies *itself* into the form of a logical system, i.e., we must suppose that a system of *physical* genera and species provides the grounds for the comparative, logical procedure of dividing natural beings according to logical genus and specific difference. This is a crucial part of the third *Critique*'s attempt to provide a transcendental justification for the application to nature of our idea of a rational, logical system, that is analogous to the first attempts in the *Critique* and in the *Metaphysical Foundations* to provide a grounding for the application to nature of our conception of a mathematical system.

3. Ak. 1, 229–30; Ak. 2, 138; Ak. 5, 400.

4. Kant is committed, in his pre-Critical works, to giving precisely this kind of explanation of order in nature in general. Accordingly, we should not think that Kant is hasty in rejecting this kind of explanation of the possibility of organisms.

5. Kant is resistant to the attempt to transform the idea of the "great chain of beings" (or the idea, in this case, that between natural objects, there are no real differences in kind but only of degree) into a physical, causal account of the variety observed in nature. His criticisms of Herder on this account are similar to Haller's

.

criticisms of Buffon. In general, even if we could make sense of matter as such successively ordering itself into organized structures, we still have not made sense of the fact that the generation of organized beings involves members of the same genera and/or species generating other members of their kind in such a way that they resemble both parents to some degree and neither exactly. Because this fact is as central to the issue as are considerations of ultimate origins, it does not appear that a purely epigenetic account of generation can be sufficient. This line of criticism is found in Haller's foreword to the German edition of Buffon's great work in natural history, with which Kant is familiar, *Allgemeine Historie der Natur nach allen ihren besonderen Theilen abgehandelt; mit einer Vorrede Herrn Doctor Albrecht von Haller, zweyter Theil* (Hamburg and Leipzig, 1752).

6.   Ak. 2, 115.

7.   The full title of this work is *Allgemeine Naturgeschichte und Theorie des Himmels oder Versuch von der Verfassung und dem mechanischen Ursprunge des ganzen Weltgebäudes, nach Newtonischen Grundsätzen abgehandelt.* The title is translated by the editors of the Cambridge edition as "Universal Natural History and Theory of the Heavens, or Essay on the Constitution and Mechanical Origin of the Entire Universe, treated in accordance with Newtonian Principles." While I don't think it a major point, it is perhaps worth mentioning that the German "allgemein" and its variants are often used to convey the sense of "general," i.e., "in abstraction from certain particularities," rather than the sense of "universal," i.e., "for all x." It is certainly not Kant's intent to explain historically each and every feature of the cosmic system, whether in accordance with Newtonian principles or not. Rather, his intent is to explain the most general features of this system historically.

8.   This view is in contrast to that of Newton, according to whom we must presuppose that God created the universe and then "turned it over" to the laws of motion. For Kant's explicit comments on his understanding of the difference between himself and Newton on this issue, see Ak. 2, 121.

9.   This position is one that Kant develops extensively in the pre-Critical works and that finds its Critical expression in the distinction between the intrinsic purposiveness we think with respect to organized beings and the merely extrinsic purposiveness thought with respect to natural arrangements that are conducive to the ends of such beings, but that are not themselves thought to be possible only as purposes.

10.   That this is, at least to some degree, Kant's intent is even more obvious in *Der einzig mögliche Beweisgrund zu einer Demonstration des Daseins Gottes* (1763, hereafter EMB). There, Kant provides a recap of the argument of the *Allgemeine Naturgeschichte* (1755, hereafter AN), as an application of his revised method for what he calls "physico-theology," but which has as much to do with suggestions for natural historical accounts of the various particular phenomena of nature as it does with using these accounts for inferring the existence of a being capable of providing the ground for the possibility of these phenomena.

11.   "Wie wäre es wohl möglich, daß Dinge von verschiedenen Naturen in Verbindung mit einander so vortreffliche Übereinstimmungen und Schönheiten zu bewirken trachten sollten, sogar zu Zwecken solcher Dinge, die sich gewissermaßen außer dem Umfange der todten Materie befinden, nämlich zum Nutzen der Menschen und Thiere, wenn sie nicht einen gemeinschaftlichen Ursprung erkennten,

nämlich einen unendlichen Verstand, in welchem aller Dinge wesentliche Beschaffenheiten beziehend entworfen worden?" (Ak. 1, 225).

"How would it even be possible that things of different natures should strive in connection with each other to effect such splendid harmonies and beauties, especially to the ends of such things as find themselves outside the sphere of dead matter, namely, for the use of humans and animals, if they did not recognize a common origin, namely, an infinite understanding, in which all things were designed with respect to their essential properties?"

12. This is the main point of the "revised method of physico-theology" suggested in *EMB*, as well as the preemptive response in the *AN* to certain theologians who will see the most convincing argument for God's existence undermined if matter is capable of ordering itself into such harmonious relationships. According to Kant, the more we are able to explain as necessary consequences following from the essence of matter, the more amazed we will be that such self-ordering matter is even possible. This will have the two-fold effect of, on the one hand, strengthening our commitment to the existence of a wise creator, and, on the other, freeing the natural sciences from any a priori strictures concerning what can be explained naturally without appeal to specific provisions.

13. One significant difference is, of course, that natural history can no longer be used in making determinate claims about the existence or characteristics of a being thought to provide the ultimate ontological ground of the natural world. Kant continues to think it inevitable that we attempt to do this, however, and revisits this issue at several points in the Critical works. Given Kant's insistence, even in the pre-Critical period, that we investigate nature as far as is possible without relying on supernatural causes, it does not appear that his giving up on the possibility of a theoretical demonstration of the existence of God has any direct bearing on his proposal for the actual practices of natural history. That is, it is not clear that a priori assurance that the natural world *actually is* the product of design would play any role with respect to the empirical investigations we make using the idea that it is as a guiding principle.

14. I have in mind here especially three essays dealing with issues of classification and the use of teleological principles, namely, "Von den verschiedenen Racen der Menschen" (1775), "Bestimmung des Begriffs einer Menschenrasse" (1785), and "Über den Gebrauch teleologischer Principien in der Philosophie" (1788).

15. Kant's lectures on physical geography are an especially valuable source for understanding the natural scientific and natural historical contexts in which Kant is writing. References made in his published texts often refer to issues of which he kept himself abreast through his preparations for these lectures. For an introduction to some of the issues Kant deals with in these lectures, as well as some of the issues Kant scholars must deal with in reconstructing these lectures, see Werner Stark, "Immanuel Kants physische Geographie—Eine Herausforderung?" A text version of Professor Stark's Inaugural Lecture at the Philipps-Universität, Marburg, is available on-line through the Marburger Kant-Archiv. I would also like to thank Professor Stark for allowing me access to the materials he is working on for a forthcoming Akademie edition of the physical geography, and Professor Reinhardt Brandt for inviting me to observe the work group devoted to this project.

16. This "showplace" can be taken to indicate both a place where marvelous and extraordinary things are exhibited and simply a place where things are seen more generally. While this distinction is not central to the points I want to make here, it is an important one concerning general attitudes toward nature in the eighteenth century. For a suggestion as to the role this distinction plays in Kant's own thinking about physical geography, see, again, Stark.

17. "Den Inbegriff aller Gegenstände nennt man Welt. Es läßt sich also auch eine Wissenschaft denken; deren Gegenstand die Welt ist d. h. die alles in sich vereiniget, und das heißt Weltkenntniß. Diese Wissenschaft wird die wichtigste und nothwendigste von allen Wissenschaften seyn, weil ohne sie alle andre nur isoliert, nicht in Verbindung ständen. Sie macht das Ganze aus; alle andre Wissenschaften sind nur theile von ihr. Alles wird in ihr mit einander verbunden.—Diese Weltkenntniß geht auf 2 Gegenstände; auf den Schauplatz der Welt und auf den Lauf der Welt. Von uns ist die Erde eine Welt, weil sie alle die Dinge in sich begreift, womit wir in Gemeinschaft stehen. Vor uns wird also auch Weltkenntniß nur so viel heißen als Erdkenntniß.—Den Schauplatz der Natur erwägt die Geographie, den Lauf der Natur die Geschichte. Beide also zusammen machen die eigentliche Weltkenntniß aus—" (Berlin, Ak.-Archiv: anonymus-Barth, "Kants Phys. Geogr. [356 S.]" NL.-Kant Nr. 14, p. 3). ("The totality of all objects is called the world. Accordingly, a science can be envisioned that has as its object the world, i.e., that unifies everything in itself and is called cognition of the world. This science will be the most important and most necessary of all sciences, since without it all others stand merely in isolation and not in connection.—This cognition of the world deals with two objects; the showplace of the world and the course of the world. According to us, the earth is a world, since it contains all things with which we stand in community. Thus, knowledge of the world will also mean the same as knowledge of the earth.—Geography is concerned with the showplace of nature, history with the course of nature. Thus, the two together constitute the authentic cognition of the world.")

"Wenn wir nun alles dasjenige, was Erfahrungen in sich enthält, Historie nennen, so werden sich 2 Theile derselben gedencken lassen.

"Beschreibungen: a) Die Geschichte desjenigen, welches zu einer und eben derselben Zeit geschiehet, dies ist die Geographie, welches nach den verschiedenen Gegenständen, wovon sie handelt, bald die physische, moralische, bald die Geographie der Gelehrsamkeit genennet wird.

"Erzählungen: b) Die Geschichte desjenigen, welches zu verschiedenen Zeiten geschehen, welches die eigentliche Historie ist, und nichts anders als eine continuation der Geographie ist, daher es zu der größten Unvollständigkeit der Historie gereichet, wenn man nicht weiß an welchem Orte eine Sache geschehen ist" (Berlin, Ak.-Archiv: anonymus-Pillau 2, "Collegium Physico Geographicum explicatum a P: Immanuel Kant. Regiomonti a: 1784 // finita a 1784 d. 1.ten Märtz [448 S.]" NL.-Kant Nr. 16, p. 2). If we call everything that experiences have in them history, then two parts of the same allow themselves to be thought.

"Descriptions a) The story of those things that happen at the same time. This is geography, which, according to the various objects which it treats, is sometimes called physical geography, sometimes moral geography, and sometimes the geography of learning.

"Accounts b) The story of those things that happen at different times. This is the authentic history, and is nothing other than the continuation of geography. Accordingly, it leads to the greatest incompleteness in history, if one does not know in what place a thing has occurred."

18.   "Wir können aber unsern Erfahrungs-Erkenntnißen eine Stelle anweisen unter den Begriffen oder nach Zeit und Raum wo sie würklich anzutreffen sind. Die Eintheilung der Erkenntniß nach Begriffen ist die logische Eintheilung; die Eintheilung nach Zeit und Raum ist die physische. Durch die logische Eintheilung wird ein systema naturae wie zE des Linnaeus; durch die physische Eintheilung wird eine geographische Naturbeschreibung zE. das Rinder-Geschlecht wird gezählt unter die vierfüßigen Thiere oder unter die mit gespaltenen Klauen. Dieses wäre eine Eintheilung in meinem Kopf also eine logische Eintheilung. Das Systema naturae ist gleichsam eine Registratur des Gantzen, da stell ich ein jedes Ding unter seinen Titel, wenn sie gleich auf der Welt in verschiedenen weit entlegenen Plätzen seyn" (Pennsylvania State University Library: Kaehler, "Collegium Physico Geographicum a Viro Excellentissimo Professore ordinario Domino Kant secundum dictata sua pertratum studio vero persecutum ab Siegismundo Kaehler Regiomonti per semestre aestivum 1775 [530 S.]," Ms. German 36, p. 9). ("We can, however, assign a place to our cognitions from experience either among our concepts or according to the time and space where they are really to be found. The division of cognition according to concepts is the logical division; the division according to time and space is the physical. Through logical division a system of nature comes about, like that of Linneaus; through the physical division there comes about a geographical natural description. For example, the genus of cattle is counted among the four-footed animals, or among those with cloven hooves. This would be a division in my head, thus a logical division. The Systema naturae is, as it were, a register of the whole, since I place each thing under its title, even if they are in different, quite remote places in the world.")

19.   For a detailed view of Linneaus' general methodology and actual procedures in coming up with his Systema naturae, see James Larson, Reason and Experience: The Representation of Natural Order in the Work of Carl von Linné (Berkeley: University of California Press, 1971). For an account of the controversies arising between the three most well-known naturalists of the eighteenth century, namely, Linneaus, Haller, and Buffon, and the influence these have on the study of nature up to Kant, see James Larson, Interpreting Nature: The Science of Living Form from Linneaus to Kant (Baltimore: Johns Hopkins University Press, 1994).

20.   "Alle unsre Erkenntniße der Welt werden geschöpft aus der Erfahrung, entweder aus eigner Erfahrung, oder aus Nachrichten von anderer Erfahrung. Diese Nachrichten sind nun entweder Beschreibungen oder Erzälungen. Iene gehören zur Geographie, diese zur Historie. Die Beschreibungen beziehen sich nun entweder auf Dinge, wie sie itzt zu gleicher Zeit sind, denn gehören sie zur neuen Geographie, oder auf Dinge, wie sie vormals zu einer und derselben Zeit gewesen sind, und denn gehören sie zur alten Geographie. Erzälungen aber müßen allemal Nachrichten von Dingen zu verschiedenen Zeit seyn, also eine Reihe von Veränderungen, die sich mit den Dingen nach einander zugetragen haben, in sich enthalten. Darin besteht der wesentliche Unterschied der Beschreibungen und Erzälungen, sowie der Geographie und Historie. Ordnung bringen wir in unsre Begriffe, wenn wir einen ieden derselben in unserm Verstande

seine Rechte Stelle anweisen. Wenn wir nun einem Dinge unter unsern Begriffen eine Stelle geben, so heist das seine logische Stelle; betrachten wir es aber nach der Stelle, die es auf der Erde selbst h<u>at</u>, so ist das seine phisische Stelle. So ist z. B. die logische Stelle aller Bäume in dem einzigen Verstandesbegriff Baum; ihre phisische Stelle aber auf der Erde so verschieden, als die Gattungen von Bäumen verschieden sind, oder vielmehr so verschieden, als die Stellen auf der Erde verschieden sind, wo dieser oder jener Baum steht. Nun nennt man ein System der Anordnung aller Dinge in der Welt nach Begriffen, ein System der Natur; eine Beschreibung aber aller Dinge in der Natur nach den Stellen welche sie wirklich auf der Erde einnehmen; heist: phisische Geographie. System der Natur und physische Geographie sind also nicht unterschieden im Obieckt, denn sie erwägen beide einerlei Gegenstände, nämlich den Schauplatz der Natur, sondern ihr Unterschied besteht blos in der Verbindung dieses Gegenstandes. Das System der natur hat nämlich blos mit der logischen Stelle der Dinge zu thun, die phisische Geographie aber allein mit der phisischen Stelle. Hieraus folgt, daß Dinge die unter derselben Gattung stehen, sich logisch verwandt sind; diese Dinge aber nur geographisch benachbart seyn können, wenn die Plätze welche sie wirklich einnehmen, nahe an einander liegen" (Barth, pp. 4–6). ("All our cognitions of the world are created from experience, either from our own experience or from reports of others' experience. Now, these reports are either descriptions or accounts. The former belong to geography, the latter to history. Descriptions refer either to things as they presently are at the same time, in which case they belong to the new geography, or to things as they previously were at one and the same time, in which case they belong to the old geography. Accounts, however, must in each case be reports of things at different times and, thus, contain a series of changes in things that have occurred one after another. Herein lies the essential difference between descriptions and accounts, just as between geography and history. We bring about order among our concepts when we assign each its proper place in our understanding. When we give a thing a place amongst our concepts, that is called a logical place; however, if we consider it according to the place that it occupies on the earth itself, that is its physical place. Thus, e.g., the logical place of all trees is in the single concept of the understanding of tree; their physical places on the earth, however, are as different as the various genera of trees are different, or better, as different as the places on the earth where this or that tree stands. Now, a system of the arrangement of all things in the world according to concepts is called a system of nature; but a description of all the things in the world according to the places that they actually occupy on the earth is called: physical geography. System of nature and physical geography are, thus, not distinguished with respect to their object, for they both consider the same object, namely, the showplace of nature. Rather, their difference lies merely in the connection of this object, namely, the system of nature has to do merely with the logical place of things, but physical geography has to do only with their physical place. It follows from this that things standing under the same genus are related to each other logically, but these things can be geographical neighbors only if the places they actually occupy are close to one another.")

21. "Was ist nun ehe, Geschichte oder Geographie? Die Geographie liegt der Geschichte zum Grunde, denn die Begebenheiten müßen sich doch worauf beziehen. Die Geschichte ist immer im Flusse aber die Dinge verändern sich und geben zu

gewißer Zeit eine gantz andere Geographie, also ist die Geographie das Substratum" (Kaehler, op cit., p. 16). ("What comes first, then, history or geography? Geography lies at the ground of history, since the events must be related to something. History is always in flux, but things change and provide completely different geographies at certain times. Thus, geography is the substratum.")

22. "Das Wort Geschichte in der Bedeutung, da es einerlei mit dem griechischem *Historia* (Erzählung, Beschreibung) ausdrückt, ist schon zu sehr und zu lange im Gebrauche, als daß man sich leicht gefallen lassen sollte, ihme eine andere Bedeutung, welche die Naturforschung des Ursprungs bezeichnen kann, zuzugestehen; zumal da es auch nicht ohne Schwierigkeit ist, ihm in der letzteren einen andern anpassenden technischen Ausdruck auszufinden" (Ak. 8, 163). ("The word 'history,' in the significance that it shares with the greek 'historia' (account, description), is already too much and too long in use to stand for granting it another significance that can indicate the natural inquiry into origins; particularly since it is also not without difficulty to find another suitable technical expression for the word in its latter significance.")

"Was aber den bezweifelten, ja gar schlechthin verworfenen Unterschied zwischen Naturbeschreibung und Naturgeschichte betrifft, so würde, wenn man unter der letzteren eine Erzählung von Naturbegebenheiten, wohin keine menschliche Vernunft reicht, z.B. das erste Entstehen der Pflanzen und Thiere, verstehen wollte, eine solche freilich, wie Hr. F. sagt, eine Wissenschaft für Götter, die gegenwärtig, oder selbst Urheber waren, und nicht für Menschen sein. Allein nur den Zusammenhang gewisser jetziger Beschaffenheiten der Naturdinge mit ihren Ursachen in der ältern Zeit nach Wirkungsgesetzen, die wir nicht erdichten, sondern aus den Kräften der Natur, wie sie sich uns jetzt darbietet, ableiten, nur blos so weit zurück verfolgen, als es die Analogie erlaubt, das wäre Naturgeschichte und zwar eine solche, die nicht allein möglich, sondern auch z.B. in den Erdtheorien (worunter des berühmten Linné seine auch ihren Platz findet) von gründlichen Naturforschern häufig genug versucht worden ist, sie mögen nun viel oder wenig damit ausgerichtet haben" (Ak. 8, 161–62). ("As for the doubted, or better completely discarded, difference between natural description and natural history, the latter would certainly be a science for gods that were present (or were themselves creators) and not for humans, as Mr. F. [Foster] says, if by natural history is understood an account of natural occurrences concerning which human reason is insufficient, e.g., the ultimate origin of plants and animals. Natural history, however, would simply be following the connection of certain current states of natural things with their causes in past time according to efficient laws that we do not invent, but derive from the forces of nature as they offer themselves to us now, only as far back as the analogy allows. Such is, further, not merely possible, but is also, for instance, in the theories of the earth (among which that of the famous Linneaus also finds its place) often enough attempted by exacting investigators of nature, regardless of how much they have thus accomplished.")

"Daher haben wir Naturbeschreibung, aber nicht Natur-Geschichte; dieser Name ist sehr falsch, den einige brauchen, und indem wir nur den Nahmen haben, so glauben wir auch die Sache zu haben, und denn denkt keiner daran an solcher Natur-Geschichte zu arbeiten. Die Geschichte der Natur enthält die Mannigfaltigkeit der Geographie, so wie sie in verschiedenen Zeiten gewesen ist, aber nicht wie es jetzt zu gleicher Zeit geschiehet, denn das ist Natur_Beschreibung; wenn ich aber die Begebenheiten der gantzen Natur wie sie zu allen Zeiten beschaffen gewesen vortrage, so liefere ich eine NaturGeschichte" (Kaehler 1774, p. 14). ("From there we have natural description, but not natural history; the latter name, which some use, is very incorrect. Since we have the name, we believe ourselves to have the thing also, and no one thinks to work on such a natural history. The history of nature contains the manifold of geography, as it was in various times, but not as it occurs now at the same time, for that is natural description. If I report on the events of the entirety of nature in its condition at all times, I deliver a natural history.")

23.   This theory is commonly associated with Malebranche and Leibniz, who are its most famous supporters among philosophers of the seventeenth century. Their reasons for subscribing to this view involve a combination of theological, metaphysical, and natural scientifc commitments, which result for both of them in the denial of real causal interaction between finite substances. Since organisms are considered to be finite substances, paradigmatically so for Leibniz, and since substances can begin only by creation, the generation of a new organism is a miracle that takes place at creation. The phenomena characterizing the gradual unfolding of this created substance, however, can be explained according to natural laws. Kant criticizes this theory on several grounds throughout his works.

24.   See Elisabeth Gasking, *Investigations into Generation 1651–1828* (Baltimore: John Hopkins University Press, 1967).

25.   Ak. 2, 114–15.

26.   For details on Haller's views in general, see Larson, *Interpreting Nature*, and Gasking, *Investigations*; for an extended discussion of his experiments concerning embryonic development in chickens, and his debate with Caspar Friedrich Wolff, see Roe, *Matter, Life and Generation*. Bonnet's views are also discussed in each of these places. For a view on the role Haller-Bonnet preformation plays in Kant's developing attitudes toward generation, see Phillip Sloan, "Preforming the Categories: Kant and Eighteenth-Century Generation Theory," *Journal of the History of Philosophy* 40 (2002): 229–53. I disagree with Sloan concerning both the development of Kant's views on generation and the role these views play with respect to Kant's general epistemological project. For present purposes, it is only the former disagreement that is relevant. Sloan claims that Kant changes his mind about the view of generation put forth in EMB and the early essay on race, a view he sees as a Haller-Bonnet type of preformation, deciding finally for an epigenetic account. As we will see in what follows, I think it makes more sense to see Kant as committed to an epigenetic account of the formation of the individual, but one that makes use of the notion of preformed germs characteristic of the species in accounting for certain features of the individual. Further, I think he continues to maintain this position at least into the *Critique of Judgment*, where he refers to the theory of *epigenesis* as a theory of *generic preformation*.

27.   That is, for both Haller and Bonnet, ovist preformation seemed to be the most attractive solution to the problem of accounting for generation. The primordia of the new individual exist in the mother's egg prior to fertilization, however, so there can be no constitutive role played by fertilization, and, accordingly, the tendency of the offspring to resemble both parents is left unaccounted for.

28.   Some proponents of the germ theory of preformation allow that the contribution of the male can alter in more fundamental ways the development of the

preformed germs. In his discussion in the Third *Critique*, Kant criticizes this attempt as appealing in an ad hoc fashion to a formative power, the denial of the possibility of which is central to their theory. See Ak. 5, 423–44.

29.   Kant makes explicit mention of both of these figures in EMB, where he first addresses in detail the generation of organic beings. See Ak. 2, 115.

30.   For a discussion of Maupertuis' role in discovering these facts, see Gasking, *Investigations*.

31.   Ak. 2, 114–15.

32.   For an interesting discussion of these changes, see Philippe Huneman's chapter 3 in this volume.

33.   I go into these issues in far greater detail in my dissertation, "Organisms and Teleology in Kant's Natural Philosophy."

5

SUCCESSION OF FUNCTIONS AND CLASSIFICATIONS IN POST-KANTIAN NATURPHILOSOPHIE AROUND 1800

STÉPHANE SCHMITT

Abstract
The concept of function plays an essential role in the classifications worked out by German naturalists at the turn of the eighteenth century. Kielmeyer (1793) talks of a gradation of five cardinal functions along the animal scale. Oken (1804) asserts a classification of animals based on a progression of sensitive functions. Here, we intend to describe the relationships of those theses with the Kantian tradition, and to show how some *Naturphilosophen*, mostly Schelling's students, assume this "teleomechanist program" (Lenoir) in their interpretations of the diversity of beings.

Introduction

The notion of vital force played an extremely important role in the life sciences in the late eighteenth and the early nineteenth century, especially in the German world. This success reveals the deep changes undergone by these sciences at that time (failure of strict mechanism, development of vitalistic conceptions, etc.). It was closely associated with the notion of function, from which it was not always distinguished very clearly. Of course, these notions shared an eminently teleological component, and this aspect was frequently discussed by scientists and philosophers. In particular, Kant's position on the value of teleological principles in life sciences was supported, challenged, or transformed in different ways by German biologists such as Blumenbach, Kielmeyer, and Reil. Many historians, such as James Larson and Timothy Lenoir, have studied this difficult question.[1]

Here we do not go into details of the philosophical implications of the development and the status of the reflection on vital forces in the history of biology, but we would rather provide some elements on a particular topic, namely, the meeting point of these problems and another crucial issue that was central in biology at that time, that is, the classification of living beings, particularly of the animals, and the use of physiological and anatomical arguments in the construction of the natural system. This is quite relevant to a study of what could be termed "Kantian biology," since classification is explicitly for Kant the focus of one kind of *Naturlehre*, exemplified by Linnaeus' natural history.[2]

We shall consider here only some German scientists. Indeed, the case of Germany is particularly interesting and specific, because of the influence of some philosophers such as Kant and Schelling, and because of the subsequent importance of German biology in the nineteenth century.

1. Vital Force

The causes of the development of the notion of vital force in the second part of the eighteenth century are manifold and complex. The quest for unity in natural sciences, and the desire to find a general law of the organic world that would be similar to the Newtonian gravitation in physics, undoubtedly played an important part. This was what led the Swiss physiologist Albrecht von Haller (1708–77) to his reflection on irritability, which, according to him, governed the behavior of living matter.[3] He clearly distinguished between sensibility (ability to feel, i.e., to transmit a received impression to soul) and irritability (ability of fibers to contract in response to a stimulus). Sensibility is bound to nerves, whereas irritability is a property of muscular fibers. More precisely, irritability is a kind of force that leads the fiber to contract when stimulated, independently of sensibility. "What therefore hinders us," he questioned, "from granting irritability to be that property of the animal gluten in the muscular fiber, such that upon being touched and provoked, it contracts, to which moreover it is unnecessary to assign any cause, just as no probable cause of attraction or gravity is assigned to matter? It is a physical cause, hidden in the intimate fabric, and discovered through experiments, which are evidence enough for demonstrating its existence, [but] which are too coarse to investigate further its cause in the fabric."[4] Thus irritability, like Newtonian gravitation, can be inferred from observed phenomena, but its ultimate cause is not accessible.

After Haller, this concept of irritability met considerable success, and many naturalists and philosophers, especially in Germany, discussed its definition and its significance in the following decades. Indeed, there was a gradual shift from the notion of irritability (as opposed to sensibility) to the more general and vague concept of "vital force" (*Lebenskraft*) or "organic force" (*organische Kraft*).

One of the directions of the theoretical thinking concerned the teleological or mechanical nature of vital forces such as irritability. The work of the physiologist Johann Friedrich Blumenbach (1752–1840), professor in Göttingen,[5] and of Immanuel Kant (especially the *Kritik der Urtheilskraft*, 1790) are the most important contributions to the study. For these authors, teleology represents a kind of tool, without which our understanding cannot conceive of living organisms. Hence it would have only a strictly regulative role, but could not lead to knowledge of living objects and their constitution. Lenoir's thesis, which has been challenged,[6] is that subsequent naturalists have no longer maintained this fundamental difference between the constitutive and the regulative value of teleology, and have established on the basis of this confusion a research program that Lenoir calls "teleomechanism," which extended as far as the mid-nineteenth century. As we have already said, we do not consider the details of this complex history. In any case, we can say that the matter is certainly more complex than Lenoir suggests and that the texts were often ambiguous on the distinction between regulative and constitutive teleology.

Another direction of the reflection, closely bound to the previous one, but distinct from it, corresponds to the progressive dismemberment of Haller's irritability in the last decades of the eighteenth century. Indeed, as early as Haller's era, some naturalists wondered about the limits of this concept and about its ability to explain the diversity of phenomena in living organisms, and they found that it was necessary to postulate many vital forces. Then they had to list them, to determine their limits, and to classify them.

For example, in his *Institutiones physiologicae* (1787), Blumenbach observed that many different kinds of observable effects existed in organisms: the formation and growth of the body, the motion of already constituted parts, and the ability of some parts to experience sensations. To explain these effects, he distinguished five vital forces, in addition to sensibility: the *nisus formativus* or *Bildungstrieb* (formative force), the most universal one, which governs the production of vital fluids and their organization into organs during development; the *vis cellulosa*, which corresponds to the contractility of the mucous tissue; the *vis muscularis*, or irritability of the muscles; the *vitae propriae* of certain organs with peculiar motion (e.g., the iris); and the *vis nervea*, or sensibility of the nervous system, which causes perception.[7]

According to Blumenbach, this order corresponds to the order of appearance of these forces in the developing body, from the *nisus formativus*, which enables the formation of the organism and its preservations throughout its life, to sensibility, which appears only after birth.

This increase in the number of vital forces opened up the way to a new research program: since all organisms did not possess all forces to the same degree, it seemed possible, and even necessary, to try to discover the logic of the distribution of these forces in the different classes of living beings and to show a relation between the systematic order and the taxonomy of vital forces.

In this respect, we have to emphasize the crucial influence of the philosopher Johann Gottfried Herder (1744–1803), particularly his book *Ideen zur Philosophie der Geschichte der Menschheit* ("Ideas for the Philosophy of the History of Mankind"), whose first part appeared in 1784 and was partly devoted to natural history.[8] Herder subscribes to a linear vision of the living world, the "Great Chain of Beings," which leads from the simplest creatures (such as minerals) to the most perfect, that is, man.[9] But he superimposed on this hierarchy of organic beings a gradation of physiological functions: for example, plants are only a "mouth," since they are wholly devoted to nutrition; then, as one follows the scale of beings, irritability becomes more important, and finally sensibility supplants the other forces.

It is very clear that this gradation primarily concerns the so-called organic forces (*organische Kräfte*) and not the structures: according to Herder, "no power in Nature is without an organ, but in no way is the organ the very power which acts by its means."[10]

This question of the delimitation and distribution of vital forces is also discussed by a former student of Blumenbach, Carl Friedrich Kielmeyer (1765–1844). This author is of major interest for our investigation because, although he published almost nothing, he enjoyed a considerable fame in Germany and even in Europe.[11] Through his lectures, which must have circulated, he had a determining (though ambiguous) influence on many subsequent naturalists. He was mentor to Georges Cuvier (1769–1832) in Stuttgart—a fact often pointed out to the reader of Cuvier's published correspondence with the German naturalist Heinrich Christian Pfaff—and he was quoted by Schelling, although he later criticized the speculations of the *Naturphilosophen*. He was familiar with some biological aspects of Kant's thinking and undoubtedly transmitted to younger *Naturforscher* these Kantian teachings, seeking to convince them of the relevance that transcendental philosophy could present for modern natural science.

In 1793, Kielmeyer delivered his famous lecture, *Über die Verhältnisse der organischen Kräfte unter einander in der Reihe der verschiedenen Organisationen, die Gesetze und Folgen dieser Verhältnisse* ("On the Relations of the Organic Forces in the Series of the Different Organizations, the Laws and Consequences of These Relations"), which formed the basis of his main published work.[12] His analysis is based on the observation that Nature is harmonious in spite of the antagonism of forces displayed by the different beings. Consequently, he wants to determine how such a complex net of forces could maintain itself, so that the natural system is preserved as a whole. He presents his aim in three points:

> First, what are the forces which are brought together in most individuals; then, what are the mutual relations of these forces in the different species of organizations, and what are the laws according to which they are altered in the series of the different organizations, and finally, how are these effects and their consequences,

namely the change and continuity of this organic world and of the species which compose it, grounded on these forces, which are their cause?[13]

Kielmeyer recognizes five forces, but these are not the same as those of Blumenbach. His list is: sensibility, irritability, the force of reproduction (including regeneration), the force of secretion, and the force of propulsion, or "capacity to bring into movement the fluids and to distribute them in the solid parts in a certain order."[14] He examines them successively (mainly the first three), and he determines their relative weight in one individual and in the different species, their distribution in the different classes, and their mutual relations. The significance of each force is estimated by observing its diversity, the frequency of its effects, and its opposition to other forces.

It is still discussed whether Kielmeyer considered those forces peculiar to organisms taken in a regulative perspective, as it is prescribed by the Kantian orthodoxy, or if Kielmayer thought that he reached the very constitution of nature by this division. Lenoir initiated a debate on the interpretation of this text—an interpretation made even more difficult by the fact that we have only various manuscript texts as sources—that has been continued by Jardine, Richards, and others.[15]

Kielmeyer portrays the laws of distribution of the forces throughout the organic series. For example, with regard to sensibility: "the manifold of possible sensations in the series of organizations decreases, as the volatility and delicacy of the remaining sensations increases within a limited field";[16] and, regarding irritability: "irritability, valued according to the permanence of its manifestations, increases, just as the speed, frequency, or variety of these manifestations and the manifold of sensations decreases."[17] We can give here a final example, regarding the force of reproduction, which, according to Kielmeyer, "valued according to the number of new individuals being formed in a determinate place, increases as the number of productive individuals, or, more generally, as the number of individuals already produced (as they appear after birth) decreases."[18] As a consequence, there is a set of inverse correlations among all these five forces, which Kielmeyer calls "compensations" (*Compensationen*).

Furthermore, Kielmeyer borrows from Blumenbach the idea of a successive apparition of the forces in the course of the individual development, and he is led to invoke the parallelism between the succession of forces in the organic series and in the development of the organism:

This ratio would thereby vary according to very simple laws in the animal series. The simplicity of these laws becomes evident when one considers that the laws according to which the organic forces are distributed among the different forms of life are exactly the same laws according to which these forces are distributed amongst individuals of the same species and even within the same individuals in different developmental stages: even men and birds are plant-like in their earliest stages of development, the reproductive force is highly excited in them during this period; at a later period the irritable element emerges in the moist substance in which they live; even the heart is possessed of almost indestructible irritability, and only later does one sense organ after another emerge, appearing almost exactly in the order of their appearance from the lowest to the highest in the series of organized beings, and what previously was irritability develops in the end into the power of understanding, or at least into its immediate material organ.[19]

This text is famous because it is often considered as one of the first expressions of the original theory of recapitulation, which was developed later by embryologists such as Meckel and Serres. Later, in a more explicitly transformist framework, this theory became Ernst Haeckel's *biogenetische Grundgesetz*, which is summarized in the sentence: "ontogeny recapitulates phylogeny" (i.e., there exists a parallelism between the developmental history of an organism and the history of its phylum). The exact meaning of recapitulation in Kielmeyer's thought can be discussed, as William Coleman and Timothy Lenoir have done,[20] but certainly Kielmeyer played a very important part in its development. Probably, the subsequent authors have not perfectly understood him when they established a structural parallelism between the developmental and systematic series. In any case, the reflection on the nonrandom distribution of forces in the different classes by Kielmeyer showed the way to the research into a classification system grounded on this repartition.

We can note that similar considerations appear in the writings of other naturalists of the time associated with what Lenoir calls "the Göttingen School." For example, Heinrich Friedrich Link (1767–1851), another student of Blumenbach and a professor at the University of Rostock, wrote a paper entitled "Über die Lebenskräfte in naturhistorischer Rücksicht" ("On the Vital Forces with Regard to Natural History") in 1795, in which he quoted Kielmeyer's classification of forces in support.[21]

2. Classifications: Forces and Organs

The notion of force is not always very clear in Kielmeyer, and it can simultaneously describe causes and effects. This weakness only expresses the general uncertainty of the concept of force in biology at that time, and its confusion with that of function. Furthermore, even if Kielmeyer's approach is eminently physiological, he says almost nothing about the relationship between organ and force, and between form and function. With Lorenz Oken, however (1779–1851), we observe a slide from force to function and from function to organ.[22] As early as 1804, the main principles of Oken's biology were set out in his *Uebersicht des Grundrisses des Sistems der Naturfilosophie und der damit entstehenden Theorie der Sinne*, a small twenty-two-page book. This was followed in the succeeding years by other texts.[23] Oken is clearly influenced by Schelling's conceptions, such as exposed in his *System der gesammten Philosophie und der Naturphilosophie insbesondere*. These lectures, written in 1804, were published only in 1860, but Oken certainly knew of these ideas when he was in Würzburg with Schelling. Schelling divides the class of animals according to the three prevailing organic functions: *Reproduktionsthiere*, "animals with reproduction" (polyps, mollusks, insects); *Irritibilitätsthiere*, "animals with irritability" (fishes, amphibians, birds); and *Sensibilitätsthiere*, "animals with sensibility" (mammals).[24] Only in this last group did Schelling classify animals according to the sense that predominates in them, and he evokes the order of their appearance in nature.

Oken went further. Above all, he wanted to build a real natural history on Schelling's *Naturphilosophie*. But he was also aware of the advances in animal systematics, especially in France. Better than Schelling, he knew of the work of the French physician Félix Vicq d'Azyr (1748–94) and of Georges Cuvier. For example, around 1770, Vicq d'Azyr in his *Discours sur l'anatomie comparée* had suggested that the animals could be classified by considering a successive addition of organs. As to Cuvier, he acknowledged the principle of subordination of characters and established a hierarchy of organs, the most central having to be considered first and foremost in the construction of animal classes. For example, the central nervous system, the heart, and the circulatory system are mentioned in order to define the characteristics of the "*embranchements*" of the animal kingdom.

Oken was well acquainted with this work, and partly drew his inspiration from it. For example, in this text from 1806, he writes:

Each animal class and each animal species is characterized by the exclusive possession of specific organs. For animals are nothing but the natural functions which have reached the highest degree of life; but, just as each of these functions has its own essence, which distinguishes it from all other functions and maintains a specific form and mode of action, it is the same for the animal or the natural animal function. Now man is the combination of all the animal characters, therefore animals are nothing but particular developments of these particular characters. As a consequence they are nothing but total representations of particular organs of man, and these organs, crystallized in them in such a pure manner, are their essence and their form. This particular organ is identical to the whole animal, whereas in man it represents only a small part.

These particular organs which have reached their totality, i.e., become a whole animal, are developed in the highest degree and in form and in action are expressed in the most pure, less mixed manner.

For all other organs are repressed, as soon as the idea of animality divides into isolated animals. On this rests the possibility of the numerous animal forms, which nevertheless do not deviate from the fundamental type; this rests on the possibility that the organs develop at the cost of the others, that the food, which had to be led to all systems, is finally given to a single one.[25]

Therefore, the different animal classes are characterized by the predominance of one function, that is, of a kind of organ, over all the others. It is precisely on this inequality of organs that any classification must necessarily be built: "Only when animal anatomy is achieved overall, and not in each particular being, can we really reach the animal system; only then we can have the idea, and according to this idea the divisions can be arranged."[26]

Like Cuvier, Oken grants a particular importance to the most central organs, namely, for him, the respiratory, digestive, and cerebral systems. Indeed, he thinks that these three systems are in fact a single one, since they share a common anatomical type, the vesicle:

If, then, the idea of animality is essentially represented under three degrees, not only the highest animal, but the whole animal kingdom must be produced according to this type; so, just as these three systems prevail in the highest animal [man], there must be an animal group where the respiratory system, with its associated organs, stands before all others, another group where it is the digestive system with its poles, and another where it is the cerebral system with its auxiliary organs. As a consequence, the animal kingdom divides into three kingdoms which are not placed side by side but one on top of the other: the kingdoms of respiration, of digestion and of cerebral action.[27]

The lower kingdom corresponds to invertebrates, the middle one to vertebrates except mammals, and the highest one to mammals:

I. Kingdom of respiration
   1. *Animalia epidermoidea* = *Vermes*
   2. *Animalia dermoidea* = *Insecta*
   3. *Animalia pneumonica* = *Conchylia*
II. Kingdom of digestion
   1. *Animalia osteoidea* = *Aves*
   2. *Animalia epatoidea* = *Pisces*
   3. *Animalia gastroidea* = *Amphibia*
III. Kingdom of cerebral system = *Mammalia*

This classification is supplemented by another one, which corresponds to the division of the five senses. So there are five classes: Dermatozoa ("animals with skin," invertebrates), Glossozoa ("animals with tongues," fish),

Rhinozoa ("animals with noses," reptiles and amphibians), Otozoa ("animals with ears," birds), and Ophthalmozoa ("animals with eyes," mammals). The division is repeated again within each class.

In fact, when we read the different texts, it may give the impression that Oken suggested a new classification each time. But these classifications are compatible. Each of them expresses an aspect of the complexity of the relationship between living beings. Oken speaks of a "stereotic net."[28] Each classification would be like a section of this three-dimensional net.

There are nevertheless some common features. One of them is the hierarchy of functions and corresponding organs. Another is recapitulation and the relation between the whole (animal kingdom, represented by man) and the parts (species). According to Oken, the human being is the "most vital" creature, who combines all senses, all functions, and all organs, as harmoniously as possible. Thus the human being summarizes the whole animal kingdom. Or, in other words, the animal kingdom is nothing but the dissociation of man:

> What is the animal kingdom other than an anatomized man, the macrozoon of the microzoon? In the former there lies open and expanded in the most beautiful order which in the latter is collected into small organs, though arranged in that same beautiful way. As the bloom takes up lovingly and intimately into itself all the parts of the plant, and so elevated in shimmering array offers them in sacrifice to Phoebus the eternally enhancing goddess of life, so man spiritualizes all of those natures that stir enclosed so miserably into the lowest fleshy vesicles and manifests a brilliant resurrection of them in himself.[29]

We find the same principles in play when Oken analyzes the morphology of a particular organism. He considers a whole series of affinities among the different parts. According to Schelling's principle of antagonism, if one can divide each body element into two parts, a passive or "terrestrial" one and an active or "solar" one, one can establish relations between all passive and all active parts in organism. This leads to surprising comparisons; for example, the "brain is the stomach or the heart of the nervous system," and "a bone is a solidified heart." This results in an original representation of the hierarchical organization of the animal body; each part of a whole is not only a component, but a compendium of this whole. So it is possible to build a kind of internal systematics of the organism, analogous to the systematics in the usual sense.[30]

Conclusion

This chapter is an introductory sketch to some aspects of the very complex question of the succession of functions in German biology around 1800.

We only wish to provide some elements in order to show the general evolution of this problem. In particular, it appears that there has been a trend toward a more and more morphological interpretation of the gradation of organic forces in the animal series. We encounter this tendency especially in Schelling's followers, such as Oken. This conscious identification of form, force, and function has many causes. It is linked to Schelling's *Naturphilosophie*, but also to the influence (possibly underestimated) of French comparative anatomy and systematics. For, even if it is often thought that Oken's biology is totally speculative and far from "positive" science, this opinion must certainly be revised. Oken knew perfectly well the work of Cuvier and other leading naturalists. His *Naturphilosophie* was embedded in the scientific programs of comparative anatomy and morphlogy of the time in France and Germany, and no unique reference to a so-called Kantian teleomechanism, or even to a misinterpretation of such a program, could account for the rise of this morphology-focused trend in science. All those reasons make it necessary to qualify the statement that a unified "Kantian program" governed German biology at that time.

Notes

1.   James L. Larson, "Vital Forces: Regulative Principles or Constitutive Agents? A Strategy in German Physiology, 1786–1802," *Isis* 70 (1979): 235–49; Timothy Lenoir, "Kant, Blumenbach, and Vital Materialism in German Biology," *Isis* 71 (1980): 77–108, and *Strategy of Life: Teleology and Mechanics in Nineteenth Century German Biology* (Dordrecht: Reidel, 1982); see also S.-E. Liedman, *Det organiska livet i tysk debatt 1795–1845* (Lund: Berlinska, 1966); Peter McLaughlin, "Blumenbach und der Bildungstrieb. Zum Verhältnis von epigenetischer Embryologie und typologischem Artbegriff," *Medizinhistorisches Journal* 17 (1982): 357–72; and Robert J. Richards, *The Romantic Conception of Life: Science and Philosophy in the Age of Goethe* (Chicago: University of Chicago Press, 2002). See Introduction of this volume.

2.   On Kant's idea of descriptive science, see Mark Fisher, chapter 4 in this volume.

3.   See Hermann Boerhaave, *Praelectiones academicae in proprias institutiones rei medicae*, ed. Albrecht von Haller, 6 vols. in 7 (Göttingen: A. Vandenhoeck, 1739–44), vol. 2, 1739, p. 129, footnote i (by Haller); A. von Haller, "De partibus corporis humani sensilibus et irritabilibus," *Commentarii Societatis Regiae Scientiarum Gottingensis*, 2, 1752 (pub. 1753), 114–58. On Haller's irritability, see, e.g., Georges Canguilhem, *La formation du concept de réflexe aux* XVIIe *et* XVIIIe *siècle* (Paris: Puf, 1955); Erich Hintzsche, "Einige kritische Bemerkungen zur Bio- und Ergographie Albrecht von Hallers," *Gesnerus* 16 (1959): 1–15; Shirley Roe, *Matter, Life, and Generation: Eighteenth Century Embryology and the Haller-Wolff Debate* (Cambridge: Cambridge University Press), 1981; and François Duchesneau, *La physiologie des Lumières. Empirisme, modèles, théories* (Den Haagen: Martinus Nijhoff, 1982); see also note 1.

4.   Albrecht von Haller, "De partibus corporis humani sensilibus et irritabilibus," 154, translated by Shirley Roe in *Matter, Life and Generation*, 33.

5.   On Blumenbach, see Timothy Lenoir, "The Göttingen School and the Development of Transcendental Naturphilosophie in the Romantic Era," *Studies in History of Biology* 5 (1981): 111–205, "Kant, Blumenbach, and Vital Materialism in German Biology," and *Strategy of Life*, ch. 1; Wolfgang Baron, "Die Anschauungen Johann Friedrich Blumenbachs über die Geschichtlichkeit der Nature," *Sudhoffs Archiv für Geschichte der Medizin und der aturwissenschaften* 47 (1963): 19–26; Peter McLaughlin, "Blumenbach und der Bildungstrieb"; and François Duchesneau, "Vitalism in Late 18th Century Physiology: The Cases of Barthez, Blumenbach, and John Hunter," in *William Hunter and the 18th-Century Medical World*, ed. William F. Bynum and Roy Porter (Cambridge: Cambridge University Press, 1985), 259–95.

6.   Kenneth L. Caneva, "Teleology with Regrets," *Annals of Science* 47 (1990): 291–300, and Richards, *Romantic Conception*.

7.   Johann Friedrich Blumenbach, *Institutiones physiologicae* (Göttingen: Dieterich, 1787), §§43–47.

8.   Johann G. Herder, *Ideen zur Philosophie der Geschichte der Menschheit* [1784], ed. B. Suphan (reprint, Hildesheim: G. Olms Verlag, 1967). On Kant and Herder, see John Zammito, *Kant, Herder and the Birth of Anthropology* (Chicago: University of Chicago Press, 2002), and Frederick Beiser, *The Fate of Reason* (Chicago: University of Chicago Press, 1990).

9.   Arthur O. Lovejoy, *The Great Chain of Being: A Study of the History of an Idea* (Cambridge, MA: Harvard University Press, 1960); and Lovejoy, "Herder: Progressionism without Transformism," in *Forerunners of Darwin: 1745–1859*, ed. Bentley Glass et al. (Baltimore: Johns Hopkins University Press, 1959), 207–21.

10.   Herder, *Ideen zur Philosophie der Geschichte der Menschheit*, V, 2, 172. On the "organic forces" in Herder and his influence on other author such as Kielmeyer, see Wolfgang Pross, "Herders und Kielmeyers Begriff der organischen Kräfte," in *Philosophie des Organismen in der Goethezeit: Studien zu Werk und Wirkung des Naturforschers Carl Friedrich Kielmeyer (1765–1844)*, ed. Kai Torsten Kanz, Boethius: Texte und Abhandlungen zur Geschichte der Mathematik und der Naturwissenschaft, 34 (Stuttgart: Steiner, 1994), 81–99, and Kai Torsten Kanz, Introduction to Carl F. Kielmeyer, *Über die Verhältnisse der organischen Kräfte untereinander, in der Reihe der verschiedenen Organisationen, die Gesetze und Folgen dieser Verhältnisse* (Marburg: Basilisken-Presse, 1993); Thomas Bach, *Biologie und Philosophie bei C. F. Kielmeyer und F. W. J. Schelling*, Schellingiana 12 (Stuttgart: Bad-Cannstatt, 2001), 150ff; and Richards, *Romantic Conception*, 222–25.

11.   On Kielmeyer's biography, see Georg-Friedrich von Jäger, "Ehrengedächtniss des Königl. Württembergischem Staatsraths von Kielmeyer," *Nova Acta Academiae Caesarae Leopoldina-Carolinae Germanicae Naturae Curiosorum* 21, no. 2 (1845): xvii–xcii; Carl Philipp Friedrich von Martius, "Carl Friedrich Kielmeyer," *Akademische Denkreden* (Leipzig, 1866): 181–209; Fritz-Heinz Holler, "Karl Friedrich Kielmeyer. Staatsrat, Professor der Naturwissenschaften, Direktor der wissenschaftlichen Sammlungen in Stuttgart. 1765–1844," *Schwäbische Lebensbilder*, 1 (1940): 313–23; William Coleman, "Kielmeyer, Carl Friedrich," in *Dictionary of Scientific Biography*. Ed. Charles Gillispie, 18 vols. (New York: Charles Scribners's Sons, 1970–90), vol. 7, pp. 366–69; and Kai Torsten Kanz, "Carl Friedrich Kielmeyer (1765–1844). Leben, Werk, Wirkung. Perspektiven der Forschung und Edition," in *Philosophie des Organismen in der Goethezeit*, 13–32. Kanz has published

a very complete bibliography of Kielmeyer's writings: *Kielmeyer-Bibliographie: Verzeichnis der Litteratur von und über den Naturforscher Carl Friedich Kielmeyer (1765–1844)*, Quellen der Wissenschaftsgeschichte, 1 (Stuttgart: Verlag für Geschichte der Naturwissenschaft und der Technik, 1991).

12.   Carl F. Kielmeyer, *Ueber die Verhältnisse der organischen Kräfte untereinander, in der Reihe der verschiedenen Organisationen, die Gesetze und Folgen dieser Verhältnisse* (Stuttgart: Academischen Buchdrukerei, 1793; reprint, Tübingen: Osiander, 1814), *Sudhoffs Archiv für Geschichte der Medizin und der Naturwissenschaften*, 23 (1930): 247–67. New edition with preface and introduction by Kai Torsten Kanz (Marburg: Basilisken-Presse, 1993).

13.   Kielmeyer, *Ueber die Verhältnisse der* organischen *Kräfte*, 7.

14.   Ibid.

15.   Lenoir, *Strategy of Life*, ch. 3; Nicholas Jardine, *Scenes of Inquiry: On the Reality of Questions in the Sciences* (Oxford: Clarendon Press, 1991); and Richards, *Romantic Conception*.

16.   *Ueber die Verhältnisse der organischen Kräfte*, 17.

17.   Ibid., 23.

18.   Ibid., 30.

19.   Ibid., 36–37.

20.   William Coleman, "Limits of the Recapitulation Theory: Carl Friedrich Kielmeyer's Critique of the Presumed Parallelism of Earth History, Ontogeny, and the Present Order of Organisms," *Isis* 64 (1973): 341–50; Lenoir, *Strategy of Life*.

21.   Heinrich Friedrich Link, "Ueber die Lebenskräfte in naturhistorischer Rücksicht," in *Beyträge zur Naturgeschichte*, vol. 2: *Ueber die Lebenskräfte in naturhistorischer Rücksicht, und die Classification der Säugthiere* (Rostock and Leipzig: Karl Christoph Stillers Buchhandlung, 1795). On Link's conception, see Stéphane Schmitt, *Les forces vitales et leur distribution dans la nature: un essai de "systématique physiologique." Textes de Kielmeyer, Link et Oken traduits et commentés* (Paris: Brepols, 2007).

22.   On Oken, see especially the biography by Alexander Ecker, *Lorenz Oken. Eine biographische Skizze. Gedächtnisrede zu dessen hundertjährigen Geburtstagsfeier gesprochen in der zweiten öffentlichen Sitzung der 52. Versammlung deutscher Naturforscher und Aerzte zu Baden-Baden am 20. September 1879 von Alexander Ecker. Durch erläuternde Zusätze und Mitteilungen aus Oken Briefwechsel vermehrt. Mit dem Portrait Oken's und einem Facsimile der Nr. 195 des I. Bandes der Isis* (Stuttgart: E. Schweizerbart, 1880); *Lorenz Oken: A biographical Dketch*, trans. Alfred Tulk (London: Kegan Paul, 1883). See also, e.g., Jean Strohl, *Lorenz Oken und Georg Büchner* (Zurich: Verlag der Corona, 1936); Rudolf Zaunick, ed., *Aus Leben und Werk von Lorenz Oken, dem Begründer der deutschen Naturforscherversammlungen* (Friburg in Brisgau, 1938); *Berichte der Naturforschenden Gesellschaft zu Freiburg im Breisgau*, special issue Oken Heft 41 (1951); Olaf Breidbach, Hans-Joachim Fliedner, and Klaus Ries, eds., *Lorenz Oken (1779–1851). Ein politischer Naturphilosoph* (Weimar: Hermann Böhlaus, 2001); and Olaf Breidbach and Michael Ghiselin, "Lorenz Oken and *Naturphilosophie* in Jena, Paris, and London," *History and Philosophy of the Life Sciences* 24 (2002): 219–47.

23.   Lorenz Oken, *Übersicht des Grundrißes des Sistems der Naturfilosophie und der damit entstehenden Theorie der Sinne* (Frankfurt am Main: P. W. Eichenberg, 1804), *Abriß der Naturphilosophie von Dr. Oken. Bestimmt zur Grundlage seiner Vorlesungen*

über Biologie (Göttingen: Vandenhoek and Ruprecht, 1805), *Die Zeugung* (Bamberg: Joseph Anton Göbhardt, 1805), and, with Dietrich Georg Kieser, *Beiträge zur vergleichenden Zoologie, Anatomie und Physiologie*, 2 vols. (Bamberg-Würzburg: Joseph Anton Göbhardt, 1806).

24. Friedrich Whilhelm von Schelling, *System der gesammten Philosophie und der Naturphilosophie insbesondere*, in *Sämmtliche Werke*, ed. Karl Friedrich Augsust Schelling (Stuttgart-Augsburg: Cotta, 1860), vol. 6, pp. 130–577.

25. Lorenz Oken, "Entwicklung der wissenschaftlichen Systematik der Thiere," in *Beiträge zur vergleichenden Zoologie, Anatomie und Physiologie*, vol. 1, pp. 103–4.

26. Ibid., 105.

27. Ibid., 107.

28. Oken, *Abriß der Naturphilosophie*, 203. On the modifications of the vision of scala naturae and the emergence of other representations (nets, maps. . .) in the second half of the eighteenth century, see Giulio Barsanti, "Buffon et l'image de la nature: de l'échelle des êtres à la carte géographique et à l'arbre généalogique," in *Buffon 88. Actes du Colloque International pour le bicentenaire de la mort de Buffon (Paris, Montbard, Dijon, 14–22 juin 1988)*, ed. Jean Gayon (Paris: Vrin, 1992), 255–96.

29. Oken, *Abriß der Naturphilosophie*, iii.

30. See Lorenz Oken, *Lehrbuch der Naturphilosophie*, 3 vols. (Iena: F. Frommann, 1809–11).

6

GOETHE'S USE OF KANT IN THE EROTICS OF NATURE

ROBERT J. RICHARDS

Abstract

Though it might provoke a throat seizure to pronounce the adjectives "erotic" and "Kantian" in the same breath, these terms best describe the bichambered heart of Goethe's morphological theory. While traveling in Italy in 1786–88, Goethe underwent transforming personal experiences that provided a deeply felt appreciation of the human form. He supposed that such appreciation depended on certain ideal structures; indeed, he even thought of these ideal structures as nature's creative instruments. Shortly after he returned to Weimar, he began a study of Kant's First *Critique*; and in 1790, he read the Third *Critique*. He resisted the first, but fell in love with the third. Kant's notions of an archetype and an archetypus intellectus provided just the theoretical structures to support his erotic inclinations; but under the tutelage of Schelling, he gave these Kantian regulative ideas a determinative use.

Introduction

By the end of July 1786, Johann Wolfgang von Goethe (1749–1832) had become mentally exhausted.[1] His duties as civil administrator to the dukedom of Carl August (1757–1828) of Saxe-Weimar-Eisenach occupied increasing amounts of time. His relationship with Charlotte von Stein (1742–1827), a sexually enticing woman of considerable intellectual agility, had dissolved into a tame companionship that emphasized for Goethe his decline into domesticity.

Ten years before, when the fiery young author of *Leiden des jungen Werthers* ("Sorrows of Young Werther," 1774) had arrived in Weimar to serve as companion to the young prince, all seemed possible. He composed plays and acted in them, carried on flirtations, and then worried about the repair of roads and the flooding of the silver mines of Illmenau, all with equal energy—the kind of energy that fueled his reckless infatuation with von

Stein, hardly to be deterred by her husband and three children. But those days had passed.

During the middle part of the 1780s, he had been working on a semi-auto-biographical novel, which like many of his other artistic projects lay as a *conceptus interruptus*. In the novel, the hero becomes involved with a troop of actors whose lives exhibit the kind of élan that was draining from Goethe's own life. In *Wilhlem Meisters theatralische Sendung* ("Wilhelm Meister's Theatrical Mission"), Mignon, an Italian adolescent of mysterious origin, sings a beautiful song of longing:

> Do you know the land where the lemon trees flower,
> Where in verdant groves the golden oranges tower?
> There a softer breeze from the deep blue heaven blows,
> The myrtle still and the lovely bay in repose.
> Do you know it?
> There! There!
> Would I go with you, O my master fair.[2]

2. Goethe's Italian Journey

Goethe himself thought of Italy as an escape from all that bore down on him in Weimar. And so on July 24, 1786, five days after the court celebrated his birthday in Carlsbad, he slipped away from his friends in the early hours of the morning, boarded a coach, and set off for the Brenner Pass. Later he sent a letter to Carl-August explaining his escape:

> The chief reason for my journey was to heal myself from the physical-moral illness from which I suffered in Germany and which made me useless; and, as well, so that I might still the burning thirst I had for true art. . . . When I first arrived in Rome, I soon realized that I really understood nothing of art and that I had admired and enjoyed only the pedestrian view of nature in works of art. Here, however, another nature, a wider field of art opened itself to me.[3]

In his later years, as Goethe reconstructed in imagination his Italienische Reise ("Italian Journey"),[4] he judged that during this period he learned what it was to be a human being. He also felt in his melancholic later years that since that time in Italy he had never again been really happy.[5]

In Rome, Goethe fell in with a colony of German artists, among whom was Johann Tischbein (1751–1829), who would render a memorable portrait of his friend in the Roman Compagne. Goethe took instruction in painting from Tischbein and from other artists, particularly, Angelika Kauffmann (1741–1807), with whom he visited the galleries and churches of Rome. During his excursions to museums, Goethe carried with him Johann Winckelmann's (1717–68)

Geschichte der Kunst des Altertums ("History of Ancient Art," 1764), and this critical work would convince him that great art could be produced only after experiencing the most sublime beauty, which the historian found perspicuously in the statuary of the ancient Greeks. Goethe took particular interest in sculpture and would cart back to Weimar many wagons filled with Greek and Roman statuary. Winckelmann believed that the experience of Greek models would instill in the artist an ineffable ideal of beauty, which would then become the guide both in the appreciation and in the execution of significant artistic pieces.[6]

Goethe traveled through the southern regions of Italy, where ancient Roman and Greek artifacts were to be found in every village. In Naples, he stayed with the English Ambassador Sir William Hamilton (1730–1803), and admired the great beauty of Hamilton's young mistress, Emma Lyon (1761–1815), who would soon become Lady Hamilton and later mistress to Lord Horatio Nelson (1758–1805). Hamilton displayed for Goethe many artifacts he had excavated from Isnearia, some fifty miles outside Naples. He amused his German friend with his treasury of erotica that came from the excavation, particularly amulets devoted to the god Priapus. In a letter to Carl August, Goethe alluded to the state of frustration to which he had been reduced by worship of this titular deity of his desires:

> The sweet, small god has relegated me to a difficult corner of the world. The public girls of pleasure are unsafe, as everywhere. The zitellen [the unmarried women] are more chaste than anywhere—they won't let themselves be touched and ask immediately, if one does something of that sort with them: e che concluderemo [and what is the understanding]? . . . What concerns the heart doesn't belong to the terminology of the present chancellery of love.[7]

In March 1787, Goethe and his artist friend Christoph Heinrich Kniep (1748–1825) sailed for Sicily. After some travel on the island they reached Palermo, and on April 16, the poet took himself to the public gardens to relax with a copy of Homer's *Odyssey*. He intended to compose a play based on the character of Nausicaa, who finds Odysseus washed up naked and filthy on the shore of her island. In Homer's version, once bathed by her attendants, magnificent Odysseus awakens keen desire in her. She takes him to her father's garden to wait while she prepares the court for his songs of adventure. Goethe regarded the public garden of Palermo more lovely than the imagined garden of Alcinous. He returned the next day to continue ruminations on his intended play. But before he knew it, as he recalled, "another spirit seized me, which had already been tailing me during these last few days." He gazed around the garden, and asked a question of the type prompted by his reading of Spinoza:

Whether I might not find the *Urpflanze* within this mass of plants? Something like that must exist! How else could I recognize that this structure or that was a plant, if they were not all formed according to a model.[8]

Goethe thus moved in imagination from contemplating the lovely Nausicaa and the glorious Odysseus in the garden of Alcinous to the provocative notion that in the real garden of Palermo he might discover a comparably beautiful and magnificent form. Even prior to his departure from Germany, Goethe had been thinking about what, in Spinozistic terms, he referred to as an "adequate idea"—an *Urtypus* that would capture the essential structure of all plants. While in his homeland, however, he had not gotten very far in working out this conception. Yet in Italy, he seems to have thought this Platonic ideal might find real embodiment.

Winckelmann had maintained that artistic talent could be honed and developed only if one had examples of sublime art, which he thought was epitomized in Greek sculpture. Recognition of the deep beauties of nature— the lived reality of the gardens of Palermo, for instance—would, accordingly, depend on the prior study of objects of great classical beauty. Goethe seemed initially persuaded of this proposition, which is why he roamed the museums and galleries of the Italian cities. Yet as he enjoyed the lakes and streams of northern Italy, traveled through the lush landscapes of southern Italy, spent time in the many gardens so wonderfully tended in Naples, Sicily, and Rome, he began to reverse the Wickelmannian principle: Goethe would come to argue that beauty in art could really be comprehended only after first experiencing beauty as a lived reality, beauty in nature. One potent experience that helped to foster that conclusion was his encounter with a young woman in Roma during his last year there.

Little is known about the girl whom Goethe met and who became his mistress, except for the descriptions found in the poem cycle he began before he returned to Weimar. In his *Erotica Romana*, as he first titled his poems—later renamed *Römische Elegien*—the poet sings of his love for a tawny skinned, dark-haired girl named Faustine. She blessedly knows nothing of Lotte and Werther; and nightly they honor the god Amor, as the poems have it: "devilishly, vigorously, and seriously."[9] The fifth elegy achieves an unsurpassed beauty, and reveals, I believe, a kind of epiphany about nature—as embodied in a love object—and a presentiment about art and ultimately science. The encounter with Faustine as expressed—or at least the scene as imagined— illustrated for Goethe the primacy of the experience of nature for the conduct of science and the execution of art. And it was this real or aesthetically constructed experience, I believe, that would lead him, after he returned to Germany, to find in Kant's *Critique of Judgment* a theory that would articulate his Italian experience and give it structure.

The fifth elegy of the *Römische Elegien* might be translated thusly:

Happy, I feel myself inspired now in this classical setting;
The ancient world and the present speak so clearly and evocatively to me.
Here I follow the advice to page through the works of the ancients,
With busy hands and daily with renewed joy.
Ah, but throughout the nights, Amor occupies me with other matters.
And if I wind up only half a scholar, I am yet doubly happy.
But do I not provide my own instruction, when I inspect the form
Of her lovely breasts, and guide my hand down her thighs?
Then I understand the marble aright for the first time: I think and compare,
And see with feeling eye, and feel with seeing hand.
Though my love steals from me a few hours of the day,
She grants me in recompense hours of the night.
We don't spend all the time kissing, but have intelligent conversation;
When sleep overcomes her, I lie by her side and think over many things.
Often I have composed poetry while in her arms, and have softly tapped out
The measure of hexameters, fingering along her spine.
In her lovely slumber, she breathes out, and I inspire
Her warm breath, which penetrates deep into my heart.
Amor trims the lamp and remembers the time
When he performed the same service for his three poets.[10]

Though a conceptual rendition of the poem strips it of aesthetic meaning, there is a kernel to be recovered. While recognizing that liability, let
me attempt one reading that coheres with much else that I have concluded
thus far about Goethe's esthetics and metaphysics, and have elaborated at
tedious length elsewhere.[11] In the first part of the poem, the poet claims to
understand great, classical art only when he can embrace its living embodiment. The white, hard marble of the statue speaks to him only after he
has experienced the brown, pliant flesh of the girl ("I inspect the form/ Of
her lively breasts, and guide my hands down her thighs"). As a result, the
visual aspects of each have taken on a new depth of tactility, and, recipro-
cally, his fingers grasp the colors of her flesh ("see with feeling eye, and feel
with seeing hand")—thus is classical art and contemporary nature synes-
thetically made one. The first part of the poem, then, suggests that the
sculpted marble and the living girl embody the same *Urtypus* of the female,
a form necessary for the actual creation of each, both by the artist and by
nature. The archetypal form expresses the kind of creative energy attributed to adequate ideas by Spinoza, whom Goethe assiduously studied before
departing for Italy. The form also serves for the aesthetic comprehension of
nature, now realized in Faustine and, consequently, in the classical statuary
of the city's museums. Here, then, is that reversal of the Winckelmannian
thesis: beauty in art, Goethe now contends, can be appreciated only after

the senses have first been schooled in the immediate experience of beauty in nature. The second part of the poem relates how this now-understood dynamical form transmutes into another artistic instantiation, this time in poetry. In an unforgettable image, the poet lies in the arms of his lover, inspiring her spirit and tapping out the hexameters of a poem he is composing—the very poem we are reading?—by counting against the vertebrae of her back. The poet is actually following the natural form of his lover in order to impress that same form on his words. Vision and feeling, reason and sense combine in love to produce an aesthetic and intuitive understanding of the unity grounding nature and art.

2. The Conclusion of Goethe's Journey

The conclusion of the Italian journey really marked a new beginning for Goethe as he returned to Weimar in the fall of 1788. The new beginning had several interrelated components. With Winckelmann, he maintained that the artist worked with an ineffable conception of beauty, whose expression would be conditioned by historical circumstance. Classical artists— whether Phidias or Homer—yet achieved a realization of the archetype of beauty in a more objective way than modern artists, a point he made in a letter to Herder. But Goethe then immediately confessed to his friend that he perhaps did not initially appreciate what Homer had wrought; but now, on his return from Sicily—his head reeling with images drawn from immediate experience of rocky coasts and stand-strewn bays—"the Odyssey for the first time has become for me a living word."[12] Just so the marbles of Polycleitos and Myron required the real experience of the lover's caress to appreciate what they had achieved. Only under the erotic authority of the living female could the forms buried in the marble come alive.

In the *Italienische Reise*, Goethe associated his remarks to Herder about the ancients' objectifying ability with a passage from a letter to von Stein and the Weimar group. The letter related what he had learned about the *Urpflanze*:

> Tell Herder that I am very near to the secret of the generation of plants and their organization. Under these skies, one can make the most beautiful observations. Tell him that I have very clearly and doubtlessly uncovered the principal point where the kernel [*Keim*] is located, and that I am in sight, on the whole, of everything else and that only a few points must yet be determined. The *Urpflanze* will be the most wonderful creation on the earth; nature herself will envy me. With this model and its key, one can, as a consequence, discover an infinity of plants—that is, even those that do not yet exist, because they could exist. It will not be some sketchy or fictive shadow or appearance, but will have an inner truth and necessity. The same law [*Gesetz*] will be applicable to all other living things.[13]

In his *Italienische Reise*, Goethe juxtaposed his observations about the objective idea used by superior artists with this passage, in which he described the objective idea—or law—used by nature.

I have argued, in these very brief reflections about Goethe's aesthetic experiences in Italy, that nature in the form of the female had a command over the poet, an erotic authority that directed him to the deeper scientific understanding of nature writ large, as well as to the comprehension of that nature found in great works of art. All of this, though, would receive more articulate form after Goethe read and adapted to his own usage Kant's Third *Critique*.

3. Goethe's Engagement with Kant

Shortly after his return to Weimar, with renewed energy and a new mistress—Christiane Volpius (1765–1816), whose figure mingles with that of Faustine in the Erotica Romana—Goethe undertook a study of Kant's First *Critique*, perhaps as an antidote to Herder's complaints about the book. He read the *Critique* assiduously and engaged Karl Leonhard Reinhold (1758–1823), the Kantian expositor at Jena, to discuss the issues in the early spring of 1789. Goethe, however, balked at the idea that in the Kantian system a veil was drawn between mind and external nature. He had the poetic conviction that the beauty of nature rushed into his eyes and gushed out of his pen. What he understood as Kantian subjectivism was simply not to his taste.

However, a year later, just after the publication of Kant's *Kritik der Urteilskraft* (1790), Goethe began a study of this new *Critique* and was delighted. His hesitating difficulties with Kant became blanketed in a cloud of enthusiasm, since the new work united his two passions: art and science. So great was his enthusiasm that during the cruel autumn of 1792, while he was accompanying his prince in retreat, as the German armies withdrew from France—with soldiers dying from disease and Goethe himself wracked with cold and hunger—he could yet engage a young schoolmaster met along the way in a discussion of the *Critique*.[14] Later, after he and Friedrich Schiller (1759–1805) had become close friends, they spent long periods in the Schiller home discussing the relationship between aesthetic judgments of beauty and teleological judgments of nature, which the *Critique* had formally united.

Goethe's thought resonated with several features of Kant's conception. The most important feature revealed by the Third *Critique*, however, was the intimate connection between aesthetic judgments about beautiful objects and scientific judgments about living creatures. Prior to reading Kant's work, Goethe had been convinced that a stealthy imagination could creep into science and inflict grave injury—at least this was the standard view that the poet-scientist nominally accepted.[15] Kant, however, had shown that creative imagination lawfully operated in both domains. The *Critique* argued that in aesthetic judgment imagination was given free play as it attempted to harmonize the purposiveness of form expressed in a work of art with natural rules for the construction of beauty—though such rules remained, according to Kant, ineffable and buried within the nature of the artist. But this consideration allowed Goethe to maintain that the nature of the artist and nature writ large had an intimate connection, as his epigram has it: "an unknown, law-like something in the object corresponds to an unknown, law-like something in the subject."[16]

The Kantian perspective also freed Goethe from the epistemological restriction that imagination had no role in science. The same artistic imagination, he came to understand, also functioned in judgments about nature. In teleological judgments about a plant or an animal, the scientists reflect on the purposive relation of anatomical parts to each other and to the whole organism. Here imagination constructs out of such relationships an *Urbild*, or archetype, an idea of the whole that displays the integral functioning of parts. Goethe himself would locate the archetype at the heart of his science of morphology, the science of his own devising that came to dominate biology in the first half of the nineteenth century and that was historicized by Darwin in the second half.[17] For Goethe, the archetype became the original type that the artist had to comprehend to render a beautiful object and that the scientist had to understand to yield that system of relationships constituting the essential structure of an organism.

Schiller's own interpretation of the Third *Critique*, as he worked it out in his great monograph *On Naïve and Sentimental Poetry*, also had a resounding impact on Goethe's conception of artistic imagination. Indeed, Goethe credited his friend's monograph with showing him that he himself was a Romantic, as he later recalled to Johann Peter Eckermann. Eckermann records him as saying:

> The concept of classical and Romantic poetry, which now has spread over the entire world and has caused so much strife and division . . . originated with me and Schiller. For poetry I had the maxim of objective experience and only wanted this to obtain. Schiller, however, argued entirely for the subjective, took his sort to be the only kind, and in order to defend against me, wrote the essay on naïve and sentimental poetry. He demonstrated to me that I myself, against my will, was a Romantic.[18]

Schiller's tract classifies and analyzes two different kinds of genius, and assumes the Kantian interpretation of that faculty. In the Third *Critique*, Kant describes genius in this way:

> Genius is the talent (natural gift) that gives the rule to art [*Kunst*]. Now since talent is an inborn, productive ability of the artist, it belongs to nature. So we might also express it this way: genius is the inborn mental trait (ingenium) through which nature gives the rule to art.[19]

The definition suggests that ineffable rules for the creation of beauty arise from the artist's nature, which is of a piece with nature writ large. Hence the artist is not cut off from nature, but comes to express the rules that nature herself has formulated for the production of beauty. The rules, though, are mute, yet effective in the free play of imagination; they become conscious only in the feeling of aesthetic pleasure that the free play induces. Kant wanted to thwart Herder's notion of artistic genius, which seemed to make artistic creativity a matter of unbridled inspiration. He sought to restrict the notion of genius by locating it in nature.[20] And this move exactly suited Goethe's own conviction that artistic freedom required objective constraint.

In light of Kant's definition, Schiller describes two kinds of genius: the artist of naïve genius, who virtually is nature—the artist who intuitively expresses without reflection the laws of nature in his work—and the sentimental genius, the artist who yet struggles with the disjunction between man's present state of disconnectedness with nature and an ideal of natural unity. Goethe quite easily fit into the first category—the artist who is nature—while Schiller thought of himself in more agonistic terms, as one who sought to reconcile an ideal of nature with the reality. Yet in his essay, Schiller suggests that some artists will unite both temperaments in one work, as for example, he says, in a book such as *Sorrow of a Young Werther.*[21] Thus Schiller showed how Goethe might be considered a poet in the Greek mode, who is virtually identified with nature, and also one of modern, sentimental—that is to say, Romantic—sensibility, who is aware of the disparity between the real and the ideal.

Goethe's direct experience of Faustine—who combined his Italian mistress and his Weimar mistress—brought him to an awareness of a three-fold relationship that Kant's *Critique* seemed to suggest. The first was the perception that the artist had to have an esthetic intuition of the *Urbild*, the archetype, of the female in order to make a judgment of beauty and to retain in imagination a standard for recreating that beauty in a sculpture, a painting, or a poem. Of course, for Goethe, the female was not simply an icon of beauty, but also a symbol of fertility, of organic creativity. Kant offered a second insight, or made vivid what Goethe already accepted, namely, that the scientist required the acquisition of the archetype through reflective judgment in order to understand the structure of particular organisms. This conviction formed the basis for Goethe's science of morphology; it meant that the esthetic intuition by which the artist comprehended the archetype would complement and lead the scientist in the analytic understanding of organisms by making the archetype a conscious *Urbild*. This might be called Goethe's principle of complementarity: namely, that artistic imagination and scientific imagination are two aspects of the same underlying critical phenomenon, that is, that artistic judgment and scientific judgment work in harmony to describe the fullness of living nature. This idea of complementarity, incidentally, is what Hermann von Helmholtz found so alluring about Goethe's contribution to science.[22] The final aspect of Goethe's own pre-Kantian assumptions that seemed confirmed by his absorbing the Third *Critique* regards the creative power of the archetype itself. In this final aspect, Goethe undoubtedly extended Kantian's position far beyond what the Könignsberg sage would have tolerated.

According to the Kantian system, we apply categories like causality and substance determinatively to create, as it were, the phenomenal realm of mechanistically interacting natural objects. But in considering biological creatures, we must initially analyze the anatomical parts in reflective search of that organizing idea that might illuminate their relationships. But Kant suggested that we could conceive of another kind of intellect, one other than ours, that might move from the intuition of the whole to that of the constituent parts instead of following our path from parts to whole. This would then be an *intellectus archtetypus*, whose very idea would be creative.[23] But this was no idle speculation, according to Kant. When the biologist attempts to understand the design of organisms, he must ultimately attribute that design to an idea of the whole—an archetype—as its cause, and then finally to a designing intellect. But this heuristic assumption, according to Kant, can never form a proposition of natural science—it must simply remain a heuristic device guiding the biologist in attempting to find mechanistic principles that might explain biological phenomena. However, in Kant's view, this search will never yield a perfectly mechanistic science of biology, and thus biology can never be an authentic science—never a *Naturwissenschaft* but only a *Naturlehre*.

Concerning this Kantian argument, Goethe made a trenchant and many-layered observation. He wrote in his collection *Zur Morphologie*:

> The author [Kant] seems here, indeed, to refer to a divine understanding. Yet, if in the moral realm we are supposed to rise to a higher region and approach the primary Being through belief in God, virtue, and immortality, then it also should be the same in the intellectual realm. We ought to be worthy, through the intuition of a continuously creative nature, of mental participation in its productivity.[24]

In this passage and in the brief essay from which it comes, Goethe attempted to muscle into philosophical acceptance a thesis similar to that of Friedrich Schelling, the young philosophical protégé who reformulated Kantianism and who turned Goethe's metaphysics to an objective idealism (not an oxymoron in the context of turn-of-the-century Jena). The thesis amounted to this: if moral experience required us to postulate God to make sense of that experience—as Kant's Second *Critique* had argued—then our experience of organisms should also require us to postulate an intellectual intuition to make sense of such experience. According to Kant, our human *Anschauungen*, intuitions, are sensible and require something be given to the mind. But we could imagine another kind of intuition, which he called intellectual. One in which the very intuition came to know its object by producing it—God's intuition, perhaps. Now, Goethe thought parity of reasoning ought to require the determinatively necessary postulation of an intellectual intuition. And this in two senses: first would be the intellectually intuitive action of nature—the assumption that nature herself, through a kind of instantiation of archetypal ideals, would create organisms according to such ideals; this was the lingering Spinozistic motif. The second interpretation that Goethe put on Kant's conception was that we also might share in this kind of intellectual intuition, presumably as the artist who creates an esthetic object and also as the scientist who penetrates the veil of nature to understand intuitively the archetypal, productive unity underlying its variegated displays. Like Schelling, whose idealism also ultimately stemmed from Kant, Goethe thus implied that if archetypal ideas were necessary for our experience of organic nature, then they had to be causal constituents of that experience—mentally creative of that experience. And there was the further implication of this analysis, namely, that in such mental creations we shared in nature's own generative power—indeed, that we become identified with nature in such activity. Goethe thus affirmed a Schellingian Spinozism: God, nature, and our intellect were one.

Of course, we might balk at Goethe's reinvention of Kant. But perhaps we ought to take our cue from Schiller, who encountered Goethe a month after his return from Italy at a dinner party. This was Schiller's first meeting with the great man, and he wrote his friend Christian Gottfried Körner about the encounter:

Yesterday [Goethe] was with us [at the Lengefelds'] and our conversation soon came to Kant. It's interesting how he dresses everything up in his own style and manner and renders rather surprisingly what he has read. But I didn't want to argue with him about matters very close to my own interests.[25]

Notes

1. This chapter extends considerations first developed in Robert J. Richards, *The Romantic Conception of Life: Science and Philosophy in the Age of Goethe* (Chicago: University of Chicago Press, 2002).

2. Johann Wolfgang von Goethe, *Wilhelm Meisters theatralische Sendung*, in *Sämtliche Werke nach Epochen seines Schaffens* (Müncher Ausgabe), ed. Karl Richter et al., 21 vols. (Munich: Carl Hanser Verlag, 1985–98), 2.2, 170. The novel was redrafted and published as *Wilhelm Meisters Lehrjahre* (1795–96).

3.   Goethe to Carl August (January 25, 1788), in *Goethes Briefe* (Hamburger Ausgabe), ed. Karl Robert Mandelkow, 4th ed., 4 vols. (Munich: C. H. Beck, 1988), vol. 2, p. 78.

4.   The first two parts of his *Italienische Reise* came out in 1816 and 1817 and the less redacted third part not until 1829. He regarded the book as a continuation of his autobiography *Dichtung und Wahrheit* ("Poetry and Truth").

5.   He confessed this to Eckermann (October 9, 1828). See Peter Eckermann, *Gespräche mit Goethe in den letzten Jahren seines Lebens*, 3rd ed. (Berlin: Aufbau-Verlag, 1987), 248–49.

6.   See, especially, Johann Joachim Winckelmann, *Geschichte der Kunst des Altertums* (1764; Darmstadt: Wissenschaftliche Buch Gesellschaft, 1972), 186–88.

7.   Goethe to Carl August (February 3, 1787), in *Goethes Briefe*, vol. 2, p. 48.

8.   Goethe, *Italienische Reise* (April 17, 1787), in *Sämtliche Werke*, 15, 327.

9.   Goethe, *Römische Elegien* (no. 4), in *Sämtliche Werke*, 3.2, 44.

10.   Goethe, *Römanische Elegien* (no. 5), in *Sämtliche Werke*, 3.2, 47.

11.   See Richards, *Romantic Conception*, 398–400.

12.   Goethe, *Italienische Reise* (May 17, 1787), in *Sämtliche Werke*, 15, 393.

13.   Goethe to Charlotte von Stein (June 8, 1787), in *Goethes Briefe*, vol. 2, p. 60.

14.   Goethe, *Campagne in Frankreich* (October 25), in *Sämtliche Werke*, 14, 439. In his discussion with the schoolmaster, Goethe indicated what he thought the central message of Kant's *Critique*: "A work of art should be treated as a work of nature, a work of nature as a work of art, and the value of each should be developed out of the intrinsic character of each, and each considered for itself alone."

15.   Goethe, "Der Versuch als Vermittler von Objekt und Subjekt," in *Sämtliche Werke*, 4.2, 326.

16.   Goethe, *Maximen und Reflexionem*, no. 1344, in *Sämtliche Werke*, 17, 942.

17.   See Robert J. Richards, *The Meaning of Evolution: The Morphological Construction and Ideological Reconstruction of Darwin's Theory* (Chicago: University of Chicago Press, 1992).

18.   Eckermann, *Gespräche mit Goethe*, p. 350 (March 21, 1830).

19.   Immanuel Kant, *Kritik der Urteilskraft* (§46), in *Immanuel Kant Werke*, ed. Wilhelm Weischedel (Wiesbaden: Insel Verlag, 1957), 5, 405–6.

20.   Zammito has argued quite persuasively that Kant had Herder in mind when he framed his definition of genius. See John Zammito, *The Genesis of Kant's Critique of Judgment* (Chicago: University of Chicago Press, 1992), 137–40.

21.   Friedrich Schiller, *Über naïve und sentimentalische Dichtung*, in *Schillers Werke*, ed. Julius Petersen et al., 55 vols. (Weimar: Böhlaus Nachfolger, 1943–), 734 n.

22.   See Richards, *The Romantic Conception*, 327–30.

23.   Kant, *Kritik der Urteilskraft*, in *Kants Werke*, 5, 526 (A346–47, B350–51).

24.   Goethe, "Anschauende Urteilskraft," *Zur Morphologie*, 1.2, in *Sämtliche Werke*, 12, 98–99.

25.   Schiller to Christian Körner (November 1, 1790), in *Schillers Werke*, 26: 54–55.

7

KANT AND BRITISH BIOSCIENCE

PHILLIP R. SLOAN

Abstract

In the British Isles, the biological theories of Kant were received within a heterodox interpretation refracted through Schelling and Coleridge. The British reception of Kant's philosophy generally went through a period of initial enthusiasm, followed by a decline of interest during the Napoleonic wars, succeeded by a revival of interest after 1814. Through the interpretations of Kant given by his British readers, a combination of Kant's and Schelling's philosophies of nature can be followed. In one remarkable synthesis of these perspectives, the British surgeon and comparative anatomist Joseph Henry Green developed in his Hunterian Lectures on Comparative Anatomy a way to combine both Lamarckian and Cuvierian perspectives and to develop a unifying picture of the history of nature that paved the way for Richard Owen's theory of the archetype. This study highlights the importance of an institutional context for the dynamic appropriation of Kantian concepts.

Introduction

Kant's reflections on issues of biology, teleology, and function, developed particularly in the second part of the *Kritik der Urteilskraft* of 1790, opened up for his contemporaries and successors several new options for the interpretation of living nature in light of the critical philosophy. If Kant had denied that there ever could be a "Newton of the grassblade," and seemed, at least until 1790, to restrict the legitimacy of genuine *Naturwissenschaft* to mathematical physics, he nonetheless offered a new way of looking at the issues of organic phenomena that transcended the categories of mathematical physics. Unintentionally, and even over his own objections, Kant became an important theorist of the "vitalist" revolution of the late eighteenth century in the life sciences taking place around him. This shift in the life sciences opened up new vistas for inquiry into embryology, physiology, the theory of disease, and the "history" of nature. It reinserted the concepts of teleology, vital causes, formal causation, and organicism into mainstream life science. A new distinction of the "organic" and "inorganic," formally codified in 1800 in Lamarck's conception of a new science of biologie, self-consciously opposed itself to the one-level ontology of inert matter and forces that for many represented the heritage of Newton and the mechanical philosophy.[1]

Several commentators, including Timothy Lenoir, James Larson, Robert Richards, and myself, have viewed Kant as an important theoretical voice in this transformation, but the exact nature of his influence and impact has been a matter of some debate. My conclusion is that Kant was seen by his contemporaries and successors in the life sciences as a genuine participant in this revolutionary shift in perspective in the sciences, but only as they abandoned his "regulative" interpretation of organicism, teleology, and special forms of causation that Kant continued to maintain in his reflections on the life sciences, and substituted for them a more constitutive understanding of these concepts that Kant would not sanction.[2]

The transition from Kant's *als ob* perspective to one in which organic beings were once again seen as realistically explained through vital agency and purposeful forces required for some only a creative misreading of Kant. For others, it depended on the reformation of the foundations of his philosophical program by his successors Fichte and, especially for the sciences, Friederich Schelling. Their reworking of systematic metaphysical foundations for some of Kant's insights allowed this new "vital" science to become something more than a flat-footed reinstitution of vitalism. As Robert Richards has recently developed in detail,[3] Schelling's mature philosophy connected inquiry into life science with a new way of envisioning the relation of thought to the world, of ideas to things, a reconceptualization of the relation of the real and ideal. For Schelling, the organic, and not the mechanical, became the overarching framework for a philosophy of nature. As he developed his insights in such works as the *Von der Weltseele, eine Hypothese der höheren Physik zur Erklärung des allgemeinen Organismus* of 1798, he sought to overcome the Kantian bifurcation between things in themselves and the world of phenomena, without embracing the transcendental idealism of Fichte (and later Hegel). Instead, he offered a dynamic philosophical science that in the life sciences guided research into embryology, natural history, comparative anatomy, and physiology, initiating in German life science a great creative moment that is only slowly being appreciated by historians of science.

My intent in this chapter is to pursue some of these issues into a non-German national context, that of the British Isles. I have selected this focus for two reasons. First, it enables us to see the creative incursion into British intellectual history of alternative programs to those of Lockean empiricism, British Newtonianism, and utilitarianism—the traditions normally emphasized in scholarship on the history of British thought and science. My concern is to develop in more detail some aspects of the "Germano-Coleridgean Doctrine," as John Stuart Mill was to term it, as a main intellectual tradition in the British Isles in the early decades of the nineteenth century.[4] I explore, through a selective examination, some of the ways in which post-Kantian German philosophical biology entered British life science.

My second concern is to deal with this transfer of ideas as a "dynamic appropriation," rather than a "static transmission" of texts.[5] To acquire this kind of understanding requires some purchase on the historical contingencies and institutional settings that were able to render an alien set of ideas and concepts—in this case those from the German tradition—relevant or attractive to British readers such that these could be appropriated in creative ways in life science.

In relating my subject to Kant and his impact on biology, I should be clear at the outset that Kant was important for these developments only indirectly, and largely through the reinterpretation, and even misinterpretation, of his thought, particularly as it was refracted for the British through Schelling and more immediately through Samuel Taylor Coleridge and a larger group of Germanophiles including such figures as William Whewell. These all give evidence of having been influenced by aspects of Kant's views, but they also combined these with endemic and other intellectual traditions to create an unusual blend of theoretical views that cannot properly be termed "Kantian."

I first review, in a summary way, aspects of the reception of German philosophy by British readers, developing some issues ignored in René Wellek's classic study of Kant's reception in England.[6] I then focus on the way in which some of Kant's insights were received and developed through one important London scientific institution, the Royal College of Surgeons of London. As a specific focus, I detail how Kant's distinction of the "history" from the "description" of nature, first set out in Kant's 1775 paper on race, and subsequently developed in his papers of the 1780s, was elaborated in this British context.

1. Appropriating German Philosophy

The British reading public learned of the work of Kant and his philosophical successors through a complex process of assimilation that has been studied in detail only partially. To gain some perspective on this, I have been conducting a review of discussions of German philosophy and science in the British Isles in the decades between 1790 and 1815 through a survey of the British periodical press and through the publication history of German works in England. Through this review, I have been able to arrive at some generalizations.

Focusing on one of the primary target periodicals in my research, the *Monthly Magazine and British Register,* provides a useful snapshot for analyzing the way in which German philosophical and scientific thought was being received in Britain. Originally founded in 1796 as a general monthly review magazine of Dissenter-Unitarian persuasion, and edited from 1796 to 1806 by the physician John Aitkin,[7] this journal was also closely associated with Thomas Beddoes and the Bristol Institute. Its initial volumes made considerable efforts to report on German science, literature, medicine, and philosophy. The emigré German physician, theoretician of dietetics, and early translator and expositor of Kantian philosophy, Anthony Florian Madinger Willich (17?–1804), is reported to have been one of its reviewers.[8] The very first volume of the *Monthly Review,* for example, contains one of the first favorable reports in English on the philosophical doctrines of Kant.[9]

But this initially favorable reception was not sustained, for a combination of intellectual and eventually political reasons. The developments in philosophy after Kant evidently suggested to the commentators of the *Monthly* a growing sectarianism in German philosophy that was criticized on several occasions. One unnamed reviewer commented that "another only possible science, Fichte's transcendental idealism, is again extolled as the queen and canon of all other sciences, and wanders in various forms and dresses in to the great bookmart at Leipsig. Let us hope that it will be more productive than its predecessors of genuine gold."[10] In 1802, the reviewer in the same journal spoke of the ascendancy of the philosophical systems of Schelling and Fichte in the German university system, "trumpeted forth as the most sublime truths" by "hundreds of youths" who accept "the most ridiculous opinions as the revelations of profound wisdom."[11] The rise of Schelling to philosophical prominence was the subject of several comments in the *Monthly* in the first decade of the century. A review of German philosophy in 1803 spoke of how the "various sects of Kantians, Fichtians, Schellingians, &c. are zealously contending with one another, and eagerly entering the lists as authors in defence of their peculiar modes of philosophising."[12] This sectarianism was commented on again in two subsequent reviews of German philosophy in 1804.[13] Schelling's disciples were likened to "an invisible church, an ideal fraternity."[14] Although Kant was the subject of a very sympathetic obituary in 1805,[15] the 1806 report on the Leipzig book fair by the unnamed reviewer commented that "Kant is not hurled or fallen suddenly from this throne, but is gradually sinking into oblivion, so that his name is now almost as seldom mentioned as that of the *Summus Aristoteles.* The difficulty, and even the danger, attending the study of the higher branches of philosophy and metaphysics, has become more and more evident; and the number of metaphysical writers seems to have again decreased in Germany."[16] By 1808, as a consequence in part of the restrictions on contacts imposed by the Continental System, extended reports on German literature and thought had dropped dramatically.

A similar situation holds if we narrow this focus to the periodical literature devoted to the life and medical sciences. Although interchange between students attending Scottish medical schools, London Hospitals, and German universities—at least until the Continental System's establishment in 1806—had provided important personal interconnections and sources of cross-fertilization between the Isles and Germany, this did not translate into evidence of strong interest in the Isles in German medical developments. The *Medical and Physical Journal*, which began in 1799 under the co-editorship of the same Anthony Willich, whom we have mentioned previously as the translator and expositor of Kant, initially made a strong effort to transmit the latest in German medicine and life science to the British. The early volumes included discussions of the work of Hufeland, accounts of the controversial medical theories of the Scots physician John Brown, who was at the time more popular in the German states than in the British Isles, and reports on the views of other German medical scientists. But over time this journal showed a pronounced shift of interest away from German sources, with increasing discussions of the teachings and practices of the great French clinicians, pathological anatomists, and surgeons of the period: Bichat, Desault, Laennec, and Corvisart. By 1813, discussion of German developments in medicine had generally ceased, and French theoretical medicine was clearly of greater interest for the professional readers.

One prominent German biomedical theorist alone seems to have maintained a prominence in British circles in the first decades of the nineteenth century, the Göttingen physician and theoretician of medicine and natural history Johann Blumenbach. His primary works were quickly translated into English.[17] Blumenbach is also one of only two honorary German members of the London Medical and Chirurgical Society listed in 1813.[18] Even in Blumenbach's case, some reviewers of his works viewed them as erroneous on many points and pleaded for more up-to-date domestic works on physiology.[19]

It would be incorrect to draw from this snapshot that Germanophilia was absent in British intellectual life after 1810. Since his return from his travels in the German states in 1797–99, an excursion that included attendance at Blumenbach's medical lectures at Göttingen, Samuel Taylor Coleridge had been an enthusiast for German philosophy, literature, and science, and he sought, unsuccessfully it seems, to share this enthusiasm.[20] Another example is the essayist Henry Crabb Robinson, the foreign war correspondent for the *London Times* from 1807 to 1808, who later was to become a member of Coleridge's inner circle. Studying at Weimar and enrolled as a student at the University of Jena from 1803 to 1804, from which he received a diploma in 1804, and where he studied under Schelling,[21] Robinson's lengthy letters of

1802–3 extolled the superiority of German philosophy and university culture over that in England to a British readership in a short-lived periodical.[22] The development in Britain of a group of enthusiasts for the philosophies and Kant, Schelling, Schubert, Ritter, and Goethe, concerned to pursue the "romantic" sciences of chemistry, biology, and electricity, has been described in Trevore Levere's studies.[23] Nonetheless, my research indicates a significant decline of interest in German intellectual productions between 1803 and 1814, certainly exacerbated by the restriction of contacts after the Franco-Prussian alliance and the establishment of the Continental System.

This scenario changed dramatically around 1814 with the termination of the Napoleonic wars. My review of periodicals displays a remarkable new level of interest in German thought generally in British periodicals that followed the publication of Anne-Louise-Germaine de Stäel's remarkable *De L'Allemagne* in November 1813. De Stäel had arrived, manuscript in hand, in London in June 1813 as an itinerant exile from Napoleonic France in the company of one of the intellectual founders of the Romantic movement, Friederich Schlegel. Her manuscript, published in London in French in November in three volumes by the prestigious publishing house of John Murray, followed shortly by its English translation and publication, *On Germany*, introduced a British reading public to a synthesis of German thought and culture that clearly stimulated a deep interest in all things German.[24]

The attention given this work in the periodical press of the day is illustrated by lengthy and detailed essay reviews by such notables as the Scots philosophers James Macintosh and Thomas Brown, supplemented by many anonymous essay reviews in major periodicals.[25] These provide a window into the impact of this remarkable work on post-Napoleonic Britain.[26]

This new enthusiasm also brought with it a renewed appreciation for Kant, but it was a Kant read backward, particularly through the lens of Schelling and his Romantic successors. The story of Kant in England, told in the classic study by René Wellek, for this reason might better be labelled the story of Kant via Schelling in England. This was particularly true of the interpretation given to Kant by Samuel Taylor Coleridge and his London disciples.

One novel feature of this assimilation of Kant by many British intellectuals is the reinterpretation of Kant's transcendental Ideas and the regulative maxims of reason in a realistic, rather than a regulative, fashion. It was a "Platonized" Kant that resulted. Coleridge even pressed the point that an epistemological realism was Kant's true teaching as he had encountered this in Germany, with the skeptical, restrictive epistemology promulgated in his name only serving as a cover for his true doctrines.[27] With this misinterpretation, Kant was essentially made into a dynamic Platonist who believed in the mental grasp of the underlying ideal structures of the world through the faculty of the imagination.[28]

To display how these philosophical currents were then able to achieve a dynamic appropriation in British life science, I turn to the interpretations of biology and the theory of life of the London surgeon and ardent Germanophile, Joseph Henry Green (1791–1863).

2. German Life Science at the College of Surgeons

Joseph Henry Green was remarkably well situated to play an important role in the introduction of German philosophical biology into the British Isles. He is reported by his primary biographer to have studied in the Germanies, especially at Hanover and Berlin, as a young medical student from 1806 to 1809, although details on this are difficult to verify.[29] In the summer of 1817, while serving as a young lecturer and demonstrator of anatomy at Saint Thomas's hospital in London, Green met Coleridge for the first time at a dinner party given by Green at his home on Lincoln Inn Fields for the visiting German literary scholar and intimate of the Romantic circle, Johann Ludwig Tieck. The result of this meeting was an invitation to Green to come to Berlin to study with Tieck's friend, the University of Berlin philosopher and disciple of Schelling, Karl Wilhelm Ferdinand Solger (1780–1819). In the company of his wife, Green traveled to Berlin in the late summer of 1817 to undertake an intensive private tutorial with Solger in German philosophy. On his return, he entered into an intensive collaborative friendship with Coleridge that would last until Coleridge's death in 1834. This included weekly tutorials in which the two exchanged their views on German philosophy and science. It was also in this period that Coleridge commenced his public lectures in London on philosophy, including lectures on Kant and Schelling.[30] Green became the possessor of Coleridge's primary manuscripts on scientific topics and many of his philosophical writings. His two-volume synthesis of Coleridge's philosophy, *Spiritual Philosophy*, was published posthumously in 1865.[31]

Green's combination of medical expertise, his personal study of the German philosophical and scientific tradition, and his professional appointments all made him an ideal vehicle by which a dynamic appropriation of German philosophical biology was realized. First achieving public prominence by his appointment in the summer of 1823 to the Hunterian lectureship in comparative anatomy at the Royal College of Surgeons in London, a position he held until 1828. In 1830 he assumed the first chair of surgery at the new King's College London, a post he occupied until 1837. He also delivered lectures in the 1840s on the philosophy of art and anatomy to the Royal Academy of Art.[32] This unusual combination of interests reflected his more general concern to integrate science, medicine, philosophy, theology, and esthetics, a synthesis expounded in detail in his *Spiritual Philosophy*. Green's philosophical synthesis of Kant, Schelling, Solger, and Blumenbach with endemic British philosophical, biological, and medical thought represents the most creative assimilation of the German tradition I have located among the British at this time.

Green's efforts at synthesis of some of these elements first reached public hearing in his lectures on comparative anatomy delivered at the College of Surgeons in London for five consecutive spring terms between 1824 and 1828. As his domestic audience received these theoretical insights, however, they did not hear them as an exposition of Kant, Schelling, or Coleridge, but rather as the teachings of the famed London surgeon John Hunter, the iconic intellectual inspirer of the College of Surgeons. Hunterian lecturers were bound by statute to expound the teachings of John Hunter and apply these to the immense anatomical museum that formed the collection at the College of Surgeons. Unfortunately, most of Hunter's manuscripts had for the main part been lost—more accurately, they had been intentionally destroyed by Everard Home in 1822—and the content of Hunter's doctrines was generally impossible to determine.[33] As a result, an ideal "open structure" was created within which the Hunterian lecturers could appropriate alien doctrines under the authority of Hunter. At Green's hands, this became the framework in which German philosophical biology was introduced to a British audience.

Green's lectures commenced on March 30, 1824, with an audience numbering possibly as many as three hundred auditors.[34] Although he began by likening John Hunter's contributions to physiology and comparative anatomy to those of Newton and Kepler in physics,[35] the subsequent content of these lectures drew almost nothing from John Hunter. Instead, it involved a direct importation of an eclectic synthesis of doctrines from Kant, Schelling, and, it seems, from Coleridge's own synthesis of these issues, into his exposition of comparative anatomy. To illustrate this claim I focus specifically on the way in which Green developed one important issue in Kant's thought, the relationship between the "history" and "description" of nature.

Kant first introduced this distinction in the post-Critical period in the context of his important lectures on the races of man in his summer anthropology course of 1775. There it had been used to differentiate the concept of natural history, employed in a taxonomic division of nature into convenient classes (*Schuleinteilung*), from the genetic and causal history of nature, as exemplified for Kant by the writings of Buffon, which offered a "physical system for the Understanding."[36] This distinction was used in this context to differentiate the Linnean concept of a taxonomic "Variety," a subdivision of the category "Species" in the Linnean classification system, from a new concept of a "race." The latter concept was to designate a material and historical lineage of organisms belonging to one and the same stem (*Stamme*).

Kant returned to this set of distinctions briefly ten years later, again in a discussion of the issue of human races, and then elaborated on this distinction

theoretically in an important paper of 1788.[37] Kant also introduced a new termi-
nological distinction into the discussions of natural history to keep his meanings
clear. The history of nature, or *Naturgeschichte* in Kant's technical meaning of
this term, was renamed a "physiogony," and distinguished from the description of
nature (*Naturbeschreibung*), redesignated "physiography."[38]

Kant's paper of 1788 was formally written to defend the use of teleological principles in philosophy. This was in response to the challenge by naturalist and geographer Georg Forster to his 1785 race paper. In an article in
the *Teutsche Merkur* in 1786, Forster had questioned the warrant for Kant's
distinction of "history" and "description" of nature, and its use to ground the
concept of a "race." Kant's response drew on his distinctions of regulative
and constitutive concepts and on those of pure and practical reason, as he
had developed these in the First and Second Critiques, both of which had
appeared only shortly before.[39]

Admitting the impossibility of "a narration [*Erzählung*] of natural events
to which no human reason extends, for example, to the first origins of plants
and animals,"[40] Kant nonetheless defends the legitimacy of a more restricted
meaning of *Naturgeschichte*.

> It is a science concerned with the causal understanding of present phenomena
> and the explanation of these phenomena by the forces of nature, even allowing
> one to reason back in time by analogy from the present to the past. This project
> differs radically from a mere classificatory description of nature. Such a history
> of nature, would, by contrast, concern itself with investigating the connection
> between certain present properties of the things of nature and their causes [*Ursa-
> chen*] in an earlier time in accordance with causal laws that we do not invent
> but rather derive from the forces of nature [*Kräften der Natur*] as they present
> themselves to us, pursued back, however, only so far as permitted by analogy.
> Indeed, this would be of a kind of history of nature [*Naturgeschichte*] that is not
> only possible, but one which is attempted frequently enough, as, for example,
> in the theories of the earth formulated by careful [*gründlichen*] natural scientists
> [*Naturforschern*] (among which the theories of the famous Linnaeus also find
> their place). . . . This distinction lies in the nature of things; and in making this
> distinction I am demanding nothing new but instead only the careful separation
> of one activity from the other, because they are totally heterogeneous. Further,
> if the description of nature [*Naturbeschreibung*] makes its appearance as a sci-
> ence [*Wissenschaft*] in all the full splendor of a great system, history of nature
> [*Naturgeschichte*] can only offer us fragments [*Bruchstücke*] or shaky [*wankende*]
> hypotheses. But even if the history of nature [*Naturgeschichte*] can, at the pres-
> ent time (and perhaps for ever), only be presented more in outline than in a
> work of practicable science (i.e., an activity in which one might find a blank
> space already marked out for the answers to most questions), such efforts are
> not, I hope, without value. For the result of separating and presenting history
> of nature [*Naturgeschichte*] as a special science distinguishable from the descrip-

tion of nature is that one might not do something with supposed insight for one of these two kinds of investigation which properly belongs to the other. I also hope that we might become more definitely acquainted with the sphere of real knowledge [wirklichlichen Erkenntnisse] in history of nature (for we already possess some knowledge of history of nature) along with knowledge of the boundaries and principles of such knowledge lying in reason itself, according to which this knowledge might be extended in the best possible manner.[41]

In these discussions of the 1780s, Kant was attempting to navigate between two poles. On the one hand lay the claim of Forster that a historical science of nature was epistemically impossible, a "science for God alone." On the other lay the ambitious speculations by his former pupil, Johann Gottfried Herder, who put forth a realistically interpreted history of nature from its first beginnings in his Ideen of 1784–85, whose work Kant reviewed critically in the Allgemeine Literaturzeitung in February and November of 1785.[42] Herder's views on the gradual development of life from an original Urtyp over the course of a historical transformation of types could not be accepted by Kant as more than a speculative hypothesis, restricted by the limits of the critical philosophy.[43]

Although the jump in time from Kant's reflections in the 1780s to Green's lectures of the 1820s is a significant one and covers a complexity of speculations on these issues by several figures that space does not allow me to elaborate, the fact that it was primarily after 1814 that a new interest by the British in a serious reading of Kant and his German successors developed renders this temporal gap less significant than it might appear. Green was aware not only of Kant's major works, but also of his essays of the 1780s.[44] Furthermore, he was situated in a context in which the differences between a historical and a descriptive understanding of nature were immediately relevant to his charge to lecture on the massive Hunterian collection at the College of Surgeons, a collection containing at the time of his lectures over 17,000 specimens, including both contemporary and fossil forms.

Green opened his first Hunterian lecture of the five-year series in the spring of 1824 with a discussion of the general issue of vitality. Dutifully invoking the authority of John Hunter in support of his theoretical views, Green contrasted Hunter's approach to those of "ancient Physiologists" who ascribed life to "a distinct vital power or propensity to each separate part, a sort of Imaginary attendant Diety—, an Invisible agent, or Vis Vitae propria."[45] In the second lecture he attempts to work out some of the inconsistencies in Hunter's own published statements on the nature of life.[46] Green does not resolve this from doctrines of John Hunter, although his audience would think this was the source, but rather follows Kant's interpretation of Blumenbach's concept of the Bildungstrieb, set forth in Kant's Kritik der Urteilskraft. On Kant's interpretation, the Bildungstrieb provided the primary example of a dynamic teleological force that manifested itself by the lawful actions of immanent constructive powers in the purposeful activity of organic beings.[47] However, in departure from Kant, who viewed Blumenbach's principle as no more than a phenomenal, and not an explanatory, force subject to all the restrictions Kant placed on teleological principles as explanations, Green accepts this force as a genuine explanation of natural biological phenomena. The *Bildungstrieb* is a force more fundamental than matter itself. Speaking of this as a *"nisus-Formativus*—or Energy of Life—though very different from the meaning of the phrase as used by the peripatetic Physiologists of the 15th century," Green introduced Blumenbach to his audience as one whose views have helped to clear up the "shades spread before comparative anatomy and Physiology."[48]

Green's treatment of Kant's distinctions of the history and description of nature in the subsequent lectures integrates these two concepts with his notion of a dynamic physiology. Furthermore, these distinctions were now utilized as a means to resolve a specific problem in the comparative anatomy of his day: the conflict between Georges Cuvier and Jean-Baptiste Lamarck currently taking place in Parisian scientific circles over the issue of historical species transformism and the unity of type. This conflict, subsequently given wider discussion in the late 1820s and early '30s through the conflict between Cuvier's and Lamarck's *Muséum* confrere Etienne Geoffroy Saint Hilaire,[49] pitted Lamarck's concept of a serial derivation of natural forms over time against Cuvier's division of forms into fixed and disconnected *embranchements*, each containing a subordination of fixed groups with their own basic plans.

In his second lecture, Green displayed to the audience two large diagrams offering different representations of the classification of animals.[50] One displayed the division of the animals into the four main *embranchements* as these had been outlined in Cuvier's *Animal Kingdom* of 1817 (plate 1, fig. 1). Cuvier had drawn from this division of the *embranchments* the conclusion that these plans were discontinuous, with each group discriminated into nesting groups of subordinate types down to the species level. Transformation between one body plan and another was deemed impossible because of inherent anatomical restraints.

Green's reason for presenting these alternative classificatory arrangements emerged in the course of this and subsequent lectures. One way of conceiving this classification—that represented by Cuvier—"is only an abbreviated discription [*sic*][,] a sort of artificial memory—Enabling a person to retain more in this way . . . than by a seperate discription [*sic*] of each variety."[51] This "Linnean" stance he then proceeded to contrast with a second alternative, illustrated by a different diagram, which he headed "The Ascending Series of Animals." This diagram was also the subject of the third lecture of April 3. From at least three sources, the basic outlines of this diagram can be recovered. It is a remarkable diagram in the historical context, particularly

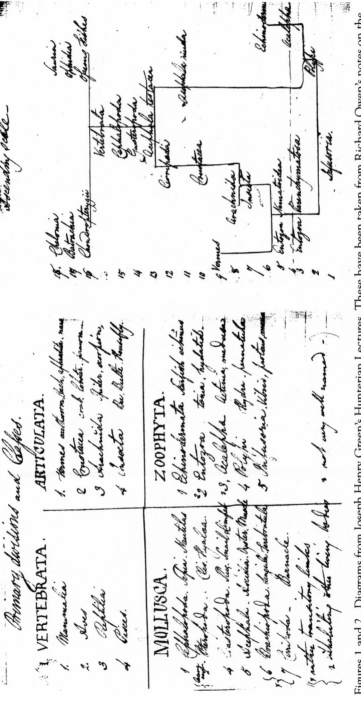

Figures 1 and 2.   Diagrams from Joseph Henry Green's Hunterian Lectures. These have been taken from Richard Owen's notes on the 1827 Lectures in the Richard Owen MSS, 275.b.21, Royal College of Surgeons. The diagrams are described but not illustrated in William Clift's notes on the Green 1824 lectures, RCS Green MSS 678.b.11. Reproduced by kind permission of the President and Council of the Royal College of Surgeons of England.

when seen in juxtaposition to the "Cuvierian" display of animal groups in the second lecture (plate 1, fig. 2).

It begins with the "simplest" forms, reversing the typical pattern of exposition of comparative anatomy that had become standard with Cuvier's lectures at the Paris museum. Green's apparent inspiration for this diagram is similar diagrams that had appeared in Jean Baptiste Lamarck's *Philosophie zoologique* and *Histoire naturelle des animaux sans vertèbres*. The curator of the museum, William Clift, records in his notes on this lecture that Green connected the two diagrams through a metaphor:

> Thus said he [with reference to the first] is all the Animal Kingdom divided—It is said he (referring to these divisions) [of the second diagram] like the Mighty Sovereign of the Forest—The Mighty oak—which from its huge trunk divides into branches—which are again and again subdivided—into lesser branches and twigs—yet from the most minute twig—all[,] all arise from the same vitality— Like these ascending gradations—Is as it were a grand March of Nature—each part in succession, shewing what will follow[.][52]

The third lecture of this initial series, delivered on April 3, made the connection with Lamarck explicit:

> I concluded my last lecture . . . with an attempt to form Animated nature into a Genesis—shall today attempt to give a general view of the gradual ascent of the Animal Kingdom—from the most simple—an ascending Series—I am exceedingly happy in this to adopt the ~~Lamarck's~~ two-fold style of Lamarck with some few alterations to my own Ideas.[53]

In common with his predecessors and successors in the Hunterian lectureship, Green had to deal with the fact that no effective guide to the massive Hunterian anatomical collection existed. Without such a key, the collection could, and was, viewed as nothing more than a miscellaneous and confusing conglomeration of bottles and anatomical preparations without discernible order. But in Green's hands, the Hunterian collection was transformed into an intelligible collection. As one contemporary reviewer commented on his opening series of lectures:

> Mr. Green's lectures at the College attracted a most crowded audience, and called forth enthusiastic plaudits, this year. The Subject was highly favourable—being no less than a delineation of the structure, functions, and natural history of the whole of animated nature, from the minutest animalcula up to man himself. The present course ascended to the conclusion of the invertebrated animals. Mr. Green not only displayed an intimate acquaintance with every link of this immense chain, but elucidated each subject by means of beautiful magnified drawings, diagrams, and figures, in aid of the costly and numerous preparations preserved in the College Museum.[54]

Green accomplished this synthesis of issues by drawing creatively on Kant's distinction of the "description" and "history" of nature. Green first seems to have presented this distinction explicitly to his audience at the opening of the 1827 lectures. "Natural history" was now defined as an enterprise that could be pursued in terms of three primary perspectives, each mutually compatible, but nonetheless distinguishable in their goals: a "physiography" or description of nature, a genetic "physiogony" or history of nature, and a "physiology" or theory of nature that connected the other two. These distinctions were then directly applied to the relations of comparative anatomy, functional physiology, and genetic history. His final statement on these relations was made in his published "Recapitulatory Lecture" that opened the 1828 series.[55]

The "physiography" of nature is concerned "with sensible experience, or appearances, in contradistinction from truths drawn from immediate facts by inference." It takes nature as product, as *natura naturata*, employing terminology derived from Spinoza, as restated in Schelling's *Naturphilosophie*.[56] On this view nature presents a series of animal groups almost as static portraits.[57] This is his interpretation of the Cuvierian comparative anatomical approach he had set forth in the first lecture of 1824.

Green then delimited a second way of analyzing nature, that as a "physiology" of nature. His use of the term "physiology" is broad and comprehensive, and has both an epistemological as well as biomedical meaning.[58] In the first sense, it is a science concerned with deducing

> by inference the rules or principles by which the innumerable facts of physiography may be reduced into manageable order, either in reference to the convenience of our faculties, which is the principle of all artificial classification, or in relation to the objects themselves, which (should it ever be realized) will be the ground of a natural classification.[59]

By this he seems to mean that rules of the relationships of natural types can be found that constitute convenient groupings for the understanding. Or this may also be a lawful grouping of things according to their real natural relationships, which in this case would constitute a natural system.

The second task of "physiology" was more in keeping with ordinary biomedical usage of the time. It is "to ascertain the powers, which must be inferred from the phaenomena, and the laws under which they act; in other words, to ascertain the idea of life and its constituent forces as far as it is common to all living bodies."[60] In other words, it is to be a science of the lawful actions of vital powers, what he had previously defined in the first lecture of 1824 as the essence of the Hunterian theory of life.

Green's interpretation of the third distinction, the physiogony or the "history" of nature, displays the concrete way in which he has assimilated Kant's definition of this science.[61] In this third view, nature is to be seen as a historical process, developing in terms of an immanent teleological purposiveness:

> The third, or Physiogony, regards the facts and appearances of the natural world as a series of actions, and Nature itself as an agent, acting under the analogy of a will and the pursuit of a purpose;—in what sense, and whether by a necessary fiction of science, or with some more substantial ground we leave here undetermined. Physiogony too, no less than physiology, investigates the principles of life; but this again principally in reference to the original construction of living bodies, and to the productive powers, or their formative principle. The distinctive aim, then, of physiogony is to present a History of Nature, and as in all other history, to discover in the past the solution of the present, and in both the anticipation of the future.[62]

This concept of a developmental history of nature, presenting "every order of living beings, from the *polypi* to the *mammalia*, as so many embryonic states of an organism, to which nature from the beginning had tended, but which Nature alone could not realize," seems to be directly behind Green's diagram of the ascending series of forms in a branching series that begins with the *infusoria*. Green describes this now as an "Idea which enabled me, in former lectures, to present to you Nature's living products, as so many significant Types of the great process which she is ever tending to complete in the evolution of the organic realm."[63]

Although Green's series of lectures maintained its primary focus on non-human life, by the conclusion of this series in 1828, his theoretical structure was brought to bear on philosophical anthropology.[64] Nature has been moving through living and extinct forms, "labouring in birth with man." Thus, "the aim of physiogony is to present the history of Nature as preface and portion of the history of man, the knowledge of Nature as a branch of self-knowledge."[65] In this view, Nature finally becomes conscious in and for itself by a process of concentration of inner powers and sensibility in specific organic forms, finally reaching the mammalian level:

> Here then we arrive at the last consummation in Nature, or rather the point in which the cycle is completed, when that which exists in itself begins to exist likewise for itself.[66]

As a final end, the process terminates in human consciousness and moral freedom:

> I have before asserted that entire intelligibility can only be given to the system of nature by an insight into an ultimate end, to which all preceding ends must be regarded as at once means and approximations,—that this ultimate end of organic nature is presented in the achievement of that sensibility, and

the subordination of the two inferior powers thereunto, by which the animal exists from itself, in itself, and, though imperfectly, for itself—and that in order to [attain] the full presentation of this ultimate end, nature must not only feel, but must know her own being. Now, this position is the same as to assert that a mind must be added to life, and consequently, that a transition from life to mind, at all events to a state in which it shall be receptive of mind, must be assumed—a transitional state, a life still retaining its essential and distinctive characters as life, but participant of mind.[67]

The principal point I wish to emphasize in this exposition is the way in which Green took Kant's distinctions and synthesized and unified these inquiries even more systematically than Kant had done himself. Green has then used these distinctions to suggest a resolution of a contemporary scientific controversy. Green also was able to use these different perspectives of a science of nature to gain some means of making sense of a concrete museum collection that previously lacked intelligibility to most viewers. Living beings are products of a constructive force of life, essentially Blumenbach's notion of the *Bildungstrieb*, which in its temporal action results in the given forms, the distinct types and species, as the historical products of its teleological drive. In this sense it provides the materials for a second enterprise, a "physiographical" understanding of nature—that is, as a description and classification of forms in the sense of Cuvierian functional analysis of discrete types and plans. But its action is also historical in the sense that it can be seen to be producing living beings in a historical process that is best understood as a branching, Lamarckian-inspired ascending series of forms in a dynamic realization of inner purpose.

In setting out selectively these intellectual components of the Green lectures, my aim has been to illustrate the way in which Green's London auditors received Kant's theoretical insights, amalgamated with views of Schelling, Blumenbach, and Coleridge, under the guise of the teachings of the revered John Hunter. They were also hearing this in a crowded lecture hall immediately attached to the massive Hunterian collection, filled not only with physiological and pathological materials, but also with fossils, whole preserved organisms, and an extensive osteological collection that they could also view. At last some kind of sense was made of this somewhat bizarre assemblage of bottles and body parts through doctrines that were being assimilated visually and audibly. As Richard Owen, who attended the 1827 and 1828 lectures of Green, was later to comment, "Every previous [Hunterian] Professor of anatomy had given some part of fragment of Zootomy in relation to his special physiological or teleological views: Mr. Green's course combined the totality, with the unity of the higher philosophy, of the Science. . . ."[68]

In supplying a German-derived comprehensive theoretical vision that pulled together both a synchronic and diachronic view of nature, Green gave the viewer of this expansive collection a theoretical rationale that it had never before possessed. Furthermore, he displayed how one could bring together opposing theoretical visions, namely the functional, anti-transformist comparative anatomy of Georges Cuvier and the developmental, historical views of Lamarck and later those of Geoffroy Saint Hilaire, who were concerned with the transformational history of life.

Kant's historical impact on British life science, we can observe, was complicated and indirect. By looking away from the German context, where most of the attention has been paid, to the reading given within another national tradition, we are able to see how aspects of Kant's project were assimilated, reformulated, and applied to new questions in creative ways. We also see in a limited way the impact of political events on this assimilation. The loss of significant contacts between the British and German communities during the Continental blockade between 1806 and 1814 provided the precondition for a dramatic resurgence of interest in German thought generally at the close of the Napoleonic wars.

This case history also highlights the importance of specific institutional contexts in shaping the "dynamics" of the appropriation of ideas. We have seen how one individual and one institution in the British Isles played a role in this assimilation. I have indicated ways in which, through this process, Kant's distinction of the history and description of nature was synthesized by a British bioscientist and applied to a complex biological problem: the relation between comparative anatomy and the derivation of forms in time. Because of the specific institutional setting in which this was done—a great anatomical museum—these philosophical abstractions were also able to obtain a concrete application to the interpretation of a collection of objects. This supplied a framework on which at least one auditor in Green's audience, Richard Owen, could begin to construct a means of relating all vertebrate forms in terms of a branching relationship.

Notes

1. For clarifying my understanding of this transformation, I am indebted to Peter Hans Reill's *Vitalizing Nature in the Enlightenment* (Berkeley: University of California Press, 2005). See also John Zammito, *Kant, Herder, and the Birth of Anthropology* (Chicago: University of Chicago Press, 2002), ch. 8; and Robert Richards, *The Romantic Conception of Life: Science and Philosophy in the Age of Goethe* (Chicago: University of Chicago Press, 2002), ch. 1.

2. For an alternative reading, see James L. Larson, "Vital Forces: Regulative Principles or Constitutive Agents? A Strategy in German Physiology, 1786–1802," *Isis* 70 (1979): 225–49.

3. Richards, *Romantic Conception*, ch. 3.

4. John Stuart Mill, "Obituary on Samuel Taylor Coleridge," *Westminster Review* (1840), in J. S. Mill, *Dissertations and Discussions*, 2 vols. (London, 1869), vol. 1, p. 403.

5. See, for this distinction, Abdelhamid I. Sabra, "The Appropriation and Subsequent Naturalization of Greek Science in Medieval Islam," *History of Science* 25 (1987): 223–43.

6. René Wellek, *Immanuel Kant in England, 1793–1838* (Princeton: Princeton University Press, 1931).

7. This journal was founded and published by Richard Phillips. For details on the history of this periodical, see Alvin Sullivan, ed., *British Literary Magazines*, vol. 2: *The Romantic Age: 1789–1836* (Westport: Greenwood Press, 1983–86). This journal included regular reviews of German philosophy, medicine, and literature in "half-year retrospects," and other information suggesting a strong interest in German developments in the early years of the journal.

8. On Willich, see Wellek, *Kant in England*, 12.

9. Thomas Beddoes, "Letter to the Editor," *Monthly Magazine and British Register* 1 (May 1796): 265–67 (hereafter MM). The three first issues also contained a review of F. A. Nitch's *A General and Introductory View of Professor Kant's Principles* (London, 1796) (2, [July–Dec. 1796], 702–5).

10. MM 12 (Aug.–Dec. 1801), 599.

11. MM 14 (Aug.–Dec. 1802), 647.

12. MM 15 (June–July 1803), 666.

13. MM 17 (Jan.–June 1804), 699–72; 18 (Aug.–Dec. 1804), 204–8.

14. MM 18 (1804), 207.

15. MM 19 (Jan.–June 1805), 354–61.

16. MM 21 (Jan.–June 1806), 550.

17. His *Ueber den Bildungstrieb* was translated in 1792. William Lawrence translated the *Outlines of Comparative Anatomy* in 1807, and his *Institutions of Physiology* was translated in 1810 by Royal College of Surgeons member John Elliotsen, with a second edition in 1815.

18. The other is Thomas Soemmering of Munich. The list of 143 members includes as foreigners Cuvier, Corvisart, Berzelius, Gaspard Vieussieux of Geneva, and Antonio Scarpa of Pavia. *London Medical and Chirurigical Transactions* 4 (1813), v–xi.

19. Anon. review of Blumenbach, *Institutions of Physiology*, *London Medical and Physical Journal* 33 (March 1815): 239–43, 243.

20. His translation of Schiller's *The Death of Wallenstein* in 1800 was reported to have fallen "dead from the press," and it was only after 1815 that his German enthusiasm seems to have been appreciated by others. See James Gillman, *The Life of Samuel Taylor Coleridge* (London: Pickering, 1838), vol. 1, pp. 146–48. His public lectures in London on philosophy, including lectures on Kant and Schelling, date from 1818.

21. Henry Crabb Robinson Papers, Williams Library (London) HCR 101.8, diploma dated December 1, 1804.

22. Henry C. Robinson, "Letters on Kant and German Literature," *Monthly Register and Encyclopedian Magazine* (1802–3); 6–12, 205–8, 294–98, 397–403. This journal lasted only from April 1802 to October 1803.

23. See especially his *Poetry Realized in Nature: Samuel Taylor Coleridge and Early Nineteenth-Century Science* (Cambridge: Cambridge University Press, 1981).

24. The publication in London of the first French edition is announced in the *London Times* on November 2, 1813. The English translation, *Germany; by the Baroness Stael-Holstein*, 3 vols. (London: J. Murray, 1814), attributed to a "Mr. Hodson," was frequently criticized for its poor quality by reviewers. On the translation, see Ghislain de Diesbach, *Madame de Staël* (Paris: Perrin, 1997), 503.

25. *Edinburgh Review* 22 (Oct. 1813): 195–238. On authorship of the reviews, see Wellek, *Kant in England*, 156. See also the review by the well-known Germanophile William Taylor, *The Monthly Review* N.S., 72 (Dec. 1813): 421–26; 73 (Jan. 1814): 63–68; (Apr. 1814): 352–65; 74 (July 1814): 268–75. On Taylor's enthusiasm for German culture, see Rosemary Ashton, *The German Idea: Four English Writers and the Reception of German Thought 1800–1860* (Cambridge: Cambridge University Press, 1980), 10–12. Other substantial unsigned essay reviews are found in *Quarterly Review* 10 (Jan. 1814): 355–409, and the *British Critic* N.S. 1 (Jan.–June 1814): 504–28, 639–59.

26. Another sign of the new enthusiasm for German culture is the increased quantity of advertisements for travel guides, German language courses, and announcement of German books by prominent booksellers advertising in the *London Times* in the 1813–14 period.

27. "In spite therefore of his own declarations, I could never believe it was possible for him to have meant no more by his *Noumenon*, or Thing in Itself, that his mere words express; or that in his own conception he confined the whole plastic power to the forms of the intellect, leaving for the external cause, for the *material* of our sensations, a matter without form, which is doubtless inconceivable." Samuel T. Coleridge, *Biographia Literaria, or Biographical Sketches of My Literary Life and Opinions* (London, 1905), 71. Such claims did not, however, meet with general agreement. The new *Blackwood's Edinburgh Magazine*, in a long essay review, treated the *Biographia* with scarcely concealed contempt, commenting that "the greatest piece of Quackery in the Book, is his pretended account of the Metaphysical System of Kant, of which he knows less than nothing" (*Blackwood's* 2 (Oct. 1817): 3–18, 17). The same periodical was, however, to be one of the main sources of favorable reviews of German literature, initiating a column in 1819 by R. P. Gillies, the "Horae Germanicae," devoted to reviewing German works. See R. Ashton, *The German Idea*, 14–15.

28. I have developed some details in this reading of Kant and the substantive interpretation of the transcendental Ideas with reference to William Whewell in my "Whewell's Philosophy of Discovery and the Archetype of the Vertebrate Skeleton: The Role of German Philosophy of Science in Richard Owen's Biology," *Annals of Science* 60 (2003): 39–61.

29. See John Simon, "Memoir of the Author's Life," in Joseph Henry Green, *Spiritual Philosophy, Founded on the Teachings of the Late Samuel Taylor Coleridge*, ed. John Simon, 2 vols. (London and Cambridge: Macmillan, 1865). For other biographical treatments, see the biography in V. G. Plarr, *Lives of the Fellows of the Royal College of Surgeons of England*, revised by d'Arcy Power et al. (London: Simpkin and Marshall, 1930), vol. 1, pp. 465–67; *DNB* 8 s.v.; and "Obituary" in *Medical Times and Gazette*, Dec. 19, 1863, pp. 650–52. The claim that Green studied for three years in Germany from 1806 to 1809 is repeated in the main biographical sources, but it has been difficult to verify in detail. Simon's account reports that he studied in both

Berlin and Hanover and that he was accompanied by his mother and on occasion joined by his father, Joseph, a British merchant. It is unusual, to say the least, that the Green family could be making such travels during the restrictions of the Continental System. I have been unable to verify Green's enrollment at the University of Göttingen or in the medical school at Göttingen in this period. This does not exclude the possibility that like many other British medical students, he may have attended the lectures of Blumenbach. Evidently, British visitors often attended Blumenbach's lectures as group. On this, see Clement Carlyon, *Early Years and Late Reflections*, 2 vols. (London: Whittaker, 1836), vol. 1, p. 187. I thank the Georg-August-Universität archivist, Dr. Haenel, and the librarian, Ms. Pfordt, for their assistance on determining more about Green's possible studies at Göttingen.

30. The first series of these lectures was delivered in the winter and spring of 1818–19.

31. On the collaboration between Green and Coleridge, see especially Heather J. Jackson, "Coleridge's Collaborator, Joseph Henry Green," *Studies in Romanticism* 21 (1982): 161–79.

32. For a summary of his later lectures on art, see *Athanaeum* 842 (Dec. 16 and 23, 1843), 1108–11, 1134–37.

33. As Green described his relation to Hunterian doctrine: "There is but one Circumstance to regret concerning John Hunter—which is that he has not left enough written documents connected with the Museum-&c- as Manuscripts of that kind would have considerably enhanced and added to the now great value of the collection." William Clift notes on Joseph Henry Green Lectures. This manuscript is headed "Lecture 7, March 30, 1824," but Green's lectures only commenced on this date and had been preceded by six lectures on surgery by Thomas Chevalier. Green MSS, Royal College of Surgeons, MS 67.b.11, fols. 21–22. Green's lectures as recorded by Clift will hereafter be cited as "Green Lectures." There are no autographs of these lectures by Green that have survived, and they must be constructed from William Clift's detailed notes. The several errors in spelling and grammar in this and other Clift transcriptions are indications of Clift's modest schooling rather than Green's errors. All manuscript quotations, and plate 1, are reproduced by kind permission of the President and Council of the Royal College of Surgeons of England.

34. The original theater at the College is reported by William Clift to have held 327 people.

35. Green Lecture 1, RCS Green MSS 67.b.11, fol. 23.

36. Kant, "Von den verschiedenen Rassen der Menschen," *Kant's Gesammelte Schriften* (Prussian Academy Edition; Berlin: Reimer, 1923), vol. 2, p. 434n. A recent English translation of the 1777 published version of this prospectus of 1775 is available in Robert Bernasconi and T. L. Lott, eds., *The Idea of Race* (Indianapolis: Hakkett, 2000), 8–22. I have discussed these issues in more detail in my "Kant on the History of Nature: The Ambiguous Heritage of the Critical Philosophy for Natural History," *Studies in History and Philosophy of the Biological and Biomedical Sciences* 37 (2006): 1–22.

37. Kant, "Bestimmung des Begriffs einer Menschenrasse" (1785), Ak. 8, 100n.; and "Ueber den Gebrauch der teleologischen Principien in der Philosophie" (1788), Ak. 8, 159–84, esp. 161–63. The first complete English translation of the 1788 paper

has been made by Jon Mark Mikkelsen in R. Bernasconi, ed., *Race* (Oxford: Blackwell, 2001), 37–56, and a translation of the 1785 paper by Mikkelsen is forthcoming. I have discussed the Kant-Forster conflict surrounding this discussion in my "Kant on the History of Nature," esp. pp. 13 ff.

38. Kant, "Ueber den Gebrauch," Ak. 8, 163n.

39. The second edition of the *Kritik der reinen Vernunft* appeared in 1787.

40. Kant, "Ueber den Gebrauch," Ak. 8, 161.

41. Ibid., 161–62, modified from the translation by J. Mikkelsen in Bernasconi, *Race*, 39.

42. Kant, "Recension," and "Erinnerungen des Recensionen der Herder'schen *Ideen zur Philosophie der Geschichte der Menschheit*," Ak. 8, 43–66. I have examined some issues in this exchange in more depth in my "Preforming the Categories: Eighteenth-Century Generation Theory and the Biological Roots of Kant's A Priori," *Journal of the History of Philosophy* 40 (2002): 229–53, esp. 242–46, and in "Kant on the History of Nature."

43. There are, to be sure, complications introduced into this picture by Kant in the well-known paragraph 80 of the *Kritik der Urteilskraft*, inspirational for Goethe and others, in which he put forth a "risky adventure" (*gewagtes Abenteuer*) of reason, open to the "archeologist" of nature, in which one could postulate the emergence of nature from chaos and the subsequent derivation of forms from a common source that are better and better adapted to circumstances. But he acknowledges that this cannot be demonstrated from experience. I conclude that this paragraph must be read against Kant's general critique of Herder's developmental pantheism and the general strictures that he has developed up to this point on the epistemic status of a developmental history of nature, which suggests that this can be for Kant no more than a speculation and never a claim to a realistic account. It was, however, read in another way by his successors. See Richards, *Romantic Conception*, esp. 443–52, and chapter 6 in this volume.

44. Green's personal copies of Kant's works are in part located in the Coleridge collection in the British Library. The sale catalog of Green's library lists Willich's translation of the *Elements of the Critical Philosophy*, three volumes of Kant's *Vermischte Schriften*, Kant's *Logik*, the *Religion innerhalb der Grenzen der blossen Vernunft*, the *Metaphysik der Sitten*, the *Sammlung Kleiner Schrifter*, and the *Anthropologie*, several listed with extensive notes by Samuel Taylor Coleridge (*Catalogue of the Library of the Late Joseph Henry Green Esq.* (London: Dryden Press, n.d.). I wish to thank Professor Heather Jackson of Victoria College of the University of Toronto for guiding me to these materials in Coleridge's library and for providing access to a copy of the Sotheby sale catalog.

45. Green lectures, RCS Green MSS 67.b.11, fol. 22. Green in fact was not following Hunter at all in these claims, who explicitly viewed life as a "superadded power." See John Hunter, "On the Vital Principle," in R. R. Palmer, ed., *The Works of John Hunter*, 4 vols. (London, 1837), vol. 1, pp. 221–28, esp. 221. This text, like most of those eventually recovered from manuscripts, many secretly copied by Clift, and published in the late 1830s and 1860s, was not available to Green. The later collection, edited by Richard Owen, was published as John Hunter, *Essays and Observations on Natural History. . .*, 2 vols. (London: Van Voorst, 1861).

46. John Hunter, "Experiments on Animals and Vegetables, with Respect to the Power of Producing Heat," *Philosophical Transactions of the Royal Society* 66 (1775): 446–58; also "Of the Heat, &c. of Animals and Vegetables," *Philosophical Transactions of the Royal Society* 68 (1778): 7–49.

47. Kant, *Kritik der Urteilskraft*, Ak. 5, 424.

48. Green, Lecture 8, April 1, 1824, RCS Green MSS 67.b.11, fol. 28.

49. See Toby Appel, *The Cuvier-Geoffroy Debate: French Biology in the Decades before Darwin* (Oxford: Oxford University Press, 1987). Lamarck had his base of operations in the main zoology gallery of the Paris Muséum National d'Histoire Naturelle and controlled the exhibits of the invertebrates. The mammals and bird were managed by Bernard de Lacépède. These exhibits, particularly those of Lamarck, were in direct opposition to the arrangements encountered by the visitor to Cuvier's Galerie d'Anatomie Comparée, also located on the same museum grounds. On this, see my "Le Muséum de Paris vient à Londres," in *Le Muséum au premier Siècle de son Histoire*, ed. Claude Blanckaert et al. (Paris: Editions du Muséum, 1997), 606–47.

50. See additional discussion in the Introduction to P. R. Sloan, ed., *Richard Owen's Hunterian Lectures, May–June 1837* (Chicago and London: University of Chicago Press/British Museum of Natural History, 1992).

51. Green Lecture 2, April 1, 1824, RCS Green MSS 67.b.11, fol. 29 (spelling and punctuation as in MS).

52. Green, Lecture 2, fol. 30.

53. Green, Lecture 3, fol. 32.

54. Unsigned reviewer, *Medical-Chirurgical Review*, N.S. 1 (1824): 251.

55. Compare with the similar language of the opening lecture of the 1827 series, published in Sloan, *Hunterian Lectures*, 306–21.

56. See Schelling, *Einleitung zu dem Entwurf eines Systems der Naturphilosophie oder über den Begriff der speculativen Physik* (1799), in *Schellings Werke*, ed. Manfred Schröter (München: Beck, 1927), vol. 2, p. 284.

57. Joseph Henry Green, "Recapitulatory Lecture, 1828," in J. H. Green, *Vital Dynamics: The Hunterian Oration Delivered in . . . February 1840* (London: Pickering, 1840), 101.

58. Compare his usage with that of Kant at the close of the first *Critique*, B 873–74 where "physiology" is defined as a primary subdivision of metaphysics, denoting the division concerned with the rational study of nature and its inner and outer connections.

59. Green, "Recapitulatory Lecture," 101.

60. "Recapitulatory Lecture," 102.

61. For a similar discussion, see Schelling, *Erster Entwurf eines Systems der Naturphilosophie* (1799). I am using the edition edited by W. Jacobs and P. Zeiche *Schelling Werke* (Stuttgart: Fromann, 2001), vol. 7, p. 116. I have not found any British author except Coleridge who made these same distinctions. On Coleridge's uses of these concepts, see Trevor Levere, *Poetry Realized in Nature*, 103–8. Levere is citing Coleridge's notebooks of 1819, and Coleridge may be the source of Green's specific way of applying these concepts.

62. "Recapitulatory Lecture," 102.

63. Ibid., 102–3.

64. Unfortunately Clift's summary of the 1828 lectures ends abruptly with a description of Green's thirteenth lecture of April 22, dealing with the senses, and does not record the extension of these themes by Green into philosophical anthropology.

65. "Recapitulatory Lecture," 103.

66. Ibid., 123.

67. Ibid., 130–31.

68. Richard Owen, Letter to John Simon, n.d., in J. Simon, "Memoir of the Author's Life," in Green, *Spiritual Philosophy*, vol. 1, pp. xiv–xv.

BIBLIOGRAPHY

Sources

Blumenbach, Johann Friedrich, and Carl Born. *Institutiones physiologicae*. Göttingen: Dieterich, 1787.

——. *Zwei Abhandlungen über die Nutritionskraft welche von der Kayserlichen Academie der Wissenschaften in Saint Petersburg den Preis gethheilt erhalten haben. Nebst einer fernern Erlaüterung eben derselben Materie von Caspar Friedrich Wolff*. St. Petersburg: Kayserlichen Academie der Wissenschaften, 1789.

Boerhaave, Hermann. *Praelectiones academicae in proprias institutiones rei medicae*. Ed. Albrecht von Haller. 6 vols. in 7. Göttingen: A. Vandenhoeck, 1739–44.

Bonnet, Charles. *Considérations sur les corps organisés, ou l'on traite de leur origine, de leur développement, de leur reproduction, etc*. Amsterdam: M. M. Rey, 1762; 2nd ed., 1768; 3rd ed., 1779. Vol. 22 of *Oeuvres d'histoire naturelle et de philosophie*. Neuchâtel: Fauche, 1779–83. Rpt., Paris: Fayard, 1985. Pp. 467–68.

——. *Lettres à Mr. l'abbé Spallanzani de Charles Bonnet*. Ed. C. Castellani. Milan: Episteme Editrice, 1971.

Buffon, Georges Louis Leclerc, Comte de. *Histoire naturelle, générale et particulière*. 3 vols. Paris: Imprimerie Royale, 1749. German translation, *Allgemeine Historie der Natur nach allen ihren besonderen Theilen abgehandelt; mit einer Vorrede Herrn Doctor Albrecht von Haller, zweyter Theil*. Hamburg and Leipzig, 1752.

Coleridge, Samuel. *Biographia Literaria, or Biographical Sketches of My Literary Life and Opinions*. London, 1905.

Eckermann, Peter. *Gespräche mit Goethe in den letzten Jahren seines Lebens*, 3rd ed. Berlin: Aufbau-Verlag, 1987.

Goethe, Johann Wolfgang von. *Goethes Briefe* (Hamburger Ausgabe), 4th ed. Ed. Karl Robert Mandelkow. 4 vols. Munich: C. H. Beck, 1988.

——. *Sämtliche Werke nach Epochen seines Schaffens*. Ed. Karl Richter et al. 21 vols. Munich: Carl Hanser Verlag, 1985–98.

Green, Joseph Henry. *Spiritual Philosophy, Founded on the Teachings of the Late Samuel Taylor Coleridge*. Ed. John Simon. 2 vols. London and Cambridge: Macmillan, 1865.

——. *Vital Dynamics: The Hunterian Oration Delivered in February 1840*. London: Pickering, 1840.

Haller, Albrecht von. *Commentarius de formatione cordis in ovo incubato*. In *Opera minora*, vol. 2 (1767). Ed. Maria Teresa Monti. Studia Halleriana VI, 54–421. Basel: Schwabe, 2000.

——. "De partibus corporis humani sensilibus et irritabilibus." *Commentarii Societatis Regiae Scientiarum Gottingensis* 2, 1752 (pub. 1753), 114–58.

——. *Elementa physiologiae corporis humani*, vol. 8. Lausanne: M. M. Bousquet,

S. d'Arnay, F. Grasset, Société Typographique, 1766.

————. *Prima linea physiologae*. Lausanne: Grasset, 1747; English translation, *First Lines of Physiology*. London, 1786.

————. "Review of *Novi Commentarii Academiae Scientiarum Imperialis Petropolitanae*, including *De formatione intestinorum* by Wolff." *Göttingische Anziegen von gelehrten Sachen* (1770): 377–81; (1771): 414–16.

————. *Sur la formation du coeur dans le poulet; sur l'oeil, sur la structure du jaune, etc.* 2 vols. Lausanne: M. M. Bousquet, 1758.

Herder, Johann Gottfried. *Ideen zur Philosophie der Geschichte der Menschheit*. Ed. B. Suphan. 1784. Rpt., Hildesheim: G. Olms Verlag, 1967.

Hunter, John. *The Works of John Hunter*. Ed. R. R. Palmer. 4 vols. London, 1837.

Kielmeyer, Carl Friedrich. *Über die Verhältnisse der organischen Kräfte untereinander, in der Reihe der verschiedenen Organisationen, die Gesetze und Folgen dieser Verhältnisse*. Ed. Kai Torsten Kanz. Marburg: Basilisken-Presse, 1993.

Link, Heinrich Friedrich. "Ueber die Lebenskräfte in naturhistorischer Rücksicht." In *Beyträge zur Naturgeschichte*, vol. 2: *Ueber die Lebenskräfte in naturhistorischer Rücksicht, und die Classification der Säugthiere*. Rostock and Leipzig: Karl Christoph Stillers Buchhandlung, 1795.

Mill, John Stuart. *Dissertations and Discussions*. 2 vols. London, 1869.

Oken, Lorenz. *Abriß der Naturphilosophie von Dr. Oken. Bestimmt zur Grundlage seiner Vorlesungen über Biologie*. Göttingen: Vandenhoek and Ruprecht, 1805.

————. *Lehrbuch der Naturphilosophie*. 3 vols. Iena: F. Frommann, 1809–11.

————. *Übersicht des Grundrißes des Sistems der Naturfilosophie und der damit entstehenden Theorie der Sinne*. Franckfurt am Main: P. W. Eichenberg, 1804.

————. *Die Zeugung*. Bamberg: Joseph Anton Göbhardt, 1805.

Oken, Lorenz, and Dietrich Georg Kieser. *Beiträge zur vergleichenden Zoologie, Anatomie und Physiologie*. 2 vols. Bamberg-Würzburg: Joseph Anton Göbhardt, 1806.

Owen, Richard. *Richard Owen's Hunterian Lectures, May-June 1837*. Ed. Phillip R. Sloan. Chicago and London: University of Chicago Press/British Museum of Natural History, 1992.

Schelling, Friedrich Whilhelm J. Von. *Erster Entwurf eines Systems der Naturphilosophie* (1799). In *Schellings Werke*. Ed. W. Jacobs and P. Zeiche. Stuttgart: Fromann, 2001.

————. *Schellings Werke*. Ed. Manfred Schröter. Munich: Beck, 1927.

————. *System der gesammten Philosophie und der Naturphilosophie insbesondere*. In *Sämmtliche Werke*. Ed. Karl Friedrich August Schelling. Stuttgart-Augsburg: Cotta, 1860.

Schiller, Fiedrich. "Über naïve und sentimentalische Dichtung." In *Schillers Werke*. Ed. Julius Petersen et al. 55 vols. Weimar: Böhlaus Nachfolger, 1943–.

Winckelmann, Johann Joachim. *Geschichte der Kunst des Altertums*. 1764. Darmstadt: Wissenschaftliche Buch Gesellschaft, 1972.

Wolff, Caspar Fridrich. "De formatione intestinorum praecipue, tum et de amnio spurio, aliisque partibus embryonis gallinacei, nondum visis." *Novi Commentarii Academiae Scientiarum Imperialis Petropolitanae* 12 (1768): 403–507; 13 (1769): 478–530. German translation by Johan Friedrich Meckel, *Über die Bildung des Darmkanals im bebrüteten Hühnchen* (Halle: Renger, 1812). French translation by

Marc Perrin, *Caspar Friedrich Wolff, la formation des intestins (1768–1769)*, with introduction and notes by Jean-Claude Dupont (Turnhout: Brepols, 2003).

———. *Theoria generationis*. Halle: Hendel, 1759. Rpt., Hildesheim: G. Olms, 1966. German translation by P. Samassa, *Ostwalds Klassiker der Exakten Wissenschaften no. 84–85* (Leipzig: Wilhelm Engelmann, 1896). Russian translation by A. E. Gaissinovitch and E. N. Pavlovski (Moscow: Edition of the Academy of Sciences of the USSR, 1950).

———. *Theorie von der Generation in zwo Abhandlungen erklärt und bewiesen*. Berlin: Friedrich Wilhelm Birnstiel, 1764. Rpt., Hildesheim: G. Olms, 1966.

———. *Von der eigenthümlichen und wesentlichen Kraft der vegetabilischen, sowohl als auch der animalischen Substanz*. St. Petersburg: Kayserliche Academie der Wissenschaften, 1789.

Studies on Kant

Adickes, Erich. *Kant als Naturforscher*. Berlin: De Gruyter, 1923.

Allison, Henry. *Kant's Transcendantal Idealism*. New Haven: Yale University Press, 2003.

Bäumler, Alfred. *Das Irrationätsproblem in der Ästhetik und Logik des 18. Jahrhunderts bis zur Kritik der Urteilskraft*. Halle: Niemeyer, 1923.

Beck, Lewis White. "Lovejoy as a Critic of Kant." In *Essays on Kant and Hume*, 61–79. New Haven: Yale University Press, 1978.

Beiser, Frederick. *The Fate of Reason*. Chicago: University of Chicago Press, 1990.

———. "Kant's Intellectual Development, 1746–1781." In *Cambridge Companion to Kant*. Ed. Paul Guyer, 26–61. Cambridge: Cambridge University Press, 1992.

———. *The Struggle against Subjectivity*. Chicago: University of Chicago Press, 2003.

Bennett, Jonathan. *Kant's Dialectic*. Cambridge: Cambridge University Press, 1974.

Bommersheim, Paul. "Der Begriff der organischen Selbstregulation in Kants *Kritik der Urteilskraft*." *Kant-Studien* 23 (1918): 209–21.

———. "Der vierfache Sinn der inneren Zweckmäßigkeit in Kants Philosophie des Organischen." *Kant-Studien* 32 (1927): 290–309.

Bowie, Andrew. *Schelling and Modern European Philosophy: An Introduction*. London: Routledge, 1993.

Brandt, Reinhardt. "Analytic/ Dialectic." In *Reading Kant*. Ed. Eva Schaper and Wilhelm Vossenkuhl. Oxford: Oxford University Press, 1989.

Buchdahl, Gerd. "Causality, Causal Laws and Scientific Theory in the Philosophy of Kant." *British Journal for Philosophy of Science* 16 (1965): 186–208.

———. "Kant's 'Special Metaphysics' and the *Metaphysical Foundations of Natural Science*." In *Kant's Philosophy of Physical Science*. Ed. Robert E. Butts, 121–61. Dordrecht: Reidel, 1986.

———. *Metaphysics and the Philosophy of Science*. Cambridge: Belknap Press, 1969.

Butts, Robert E. "Kant's Schemata as Semantics Rules." In *Kant Studies Today*. Ed. Lewis W. Beck, 290–300. Lasalle, IL: Open Court, 1969.

———. "Teleology and Scientific Method in Kant's *Critique of Judgment*." *Nous* 24 (1990): 1–16.

Bynum, William. "The Anatomical Method, Natural Theology, and the Functions of the Brain." *Isis* 64 (1973): 445–68.

Caneva, Kenneth. "Teleology with Regrets." *Annals of Science* 47 (1990): 291–300.

Cassirer, Heinrich W. *A Commentary of Kant's* Critique of Judgment. London: Methuen, 1938.

Ewing, Alfred Cyril. *Kant's Treatment of Causality*. Oxford: Archon Books, 1969.

Ferrari, Jean. "Kant lecteur de Buffon." In *Buffon 88. Actes du Colloque International pour le bicentenaire de la mort de Buffon (Paris, Montbard, Dijon, 14–22 juin 1988)*. Ed. Jean Gayon, 155–62. Paris: Vrin, 1992.

Friedman, Michael. "Causal Laws and the Foundations of Natural Science." In *Cambridge Companion to Kant*. Ed. Paul Guyer, 161–99. Cambridge: Cambridge University Press, 1992.

———. *Kant and the Exact Sciences*. Cambridge: Cambridge University Press, 1992.

Genova, Arthur. "Kant's Epigenesis of Pure Reason." *Kant-Studien* 65 (1974): 259–73.

Ginsborg, Hannah. "Kant on Understanding Organisms as Natural Purposes." In *Kant and the Sciences*. Ed. Eric Watkins, 231–59. Oxford: Oxford University Press, 2001.

———. "Two Kinds of Mechanical Inexplicability in Kant and Aristotle." *Journal of the History of Philosophy* 42, no. 1 (2004): 33–65.

Grier, Michelle. *Kant's Doctrine of Transcendental Illusion*. Cambridge: Cambridge University Press, 2001.

Guyer, Paul. "Organisms and the Unity of Science." In *Kant and the Sciences*. Ed. Eric Watkins, 259–82. Oxford: Oxford University Press, 2001.

———. "Reason and Reflective Judgment: Kant on the Significance of Systematicity," *Noûs* 24, no. 1 (1990): 17–43.

Huneman, Philippe. "Espèce et adaptation chez Kant et Buffon." In *Kant et la France-Kant und Frankreich*. Ed. Robert Theis, Jean Ferrari, and Margit Ruffin, 107–20. Hildesheim: Olms, 2005.

———. "From Comparative Anatomy to the 'Adventures of Reason.'" *Studies in History and Philosophy of Biological and Biomedical Sciences* 37, 4, (2006): 627–48.

———. "From the *Critique of Judgement* to the Hermeneutics of Nature: Sketching the Fate of the Philosophy of Nature after Kant." *Continental Philosophy Review* (2006): 1–34.

———. "Kant's Critique of the Leibnizian Theory of Organisms: An Unnoticed Cornerstone for Criticism?" *Yeditepe'de Felsefe (Istanbul Philosophical Review)* 4 (2005): 114–50.

———. *Métaphysique et Biologie. Kant et la constitution du concept d'organisme*. Paris: Puf, forthcoming.

Ingensiep, Hans."Die biologischen Analogien und die erkenntnistheoretischen Alternativen in Kants *Kritik der reinen Vernunft* B §27." *Kant-Studien* 85 (1994): 381–93.

Kemp-Smith, Norman. *A Commentary to Kant's* Critique of Pure Reason. London: Macmillan, 1923.

Kitcher, Philip. "Projecting the Order of Nature." In *Kant's Philosophy of Physical Science*. Ed. Robert E. Butts, 201–35. Dordrecht: Reidel, 1986.

Kuehn, Manfred. "Kant's masters in the exact sciences." *Kant and the Sciences*. Ed. Eric Watkins, 11–31. Oxford: Oxford University Press.

Lebrun, Gérard. *Kant et la fin de la métaphysique*. Paris: Armand Colin, 1970.

Lieber, Hans Jorg. "Kants Philosophie des Organischen und die Biologie seiner Zeit." *Philosophia Naturalis* 1 (1950): 553–70.

Longuenesse, Béatrice. *Kant et le pouvoir de juger*. Paris: Puf, 1993.

Lovejoy, Artthur. *The Great Chain of Being: A Study of the History of an Idea*. Cambridge, MA: Harvard University Press, 1960.

———. "Kant and Evolution." In *Forerunners of Darwin, 1745–1859*. Ed. Bentley Glass, Oswei Temkin, and William Strauss, 173–206. Baltimore: Johns Hopkins University Press, 1959.

Löw, Reinhard. *Philosophie des Lebendigen. Der Begriff der Organischen bei Kant, sein Grund und seine Aktualität*. Frankfort: Suhrkamp, 1980.

MacFarland, John. *Kant's Concept of Teleology*. Edinburg: Edinburg University Press, 1970.

Mahieu, Vittorio. *L'opus postumum di Kant*. Naples: Bibliopolis, 1991.

Makkreel, Rudolf. "Kant on the Scientific Status of Psychology, Anthropology and History." In *Kant and the Sciences*. Ed. Eric Watkins, 185–201. Oxford: Oxford University Press, 2001.

Marc-Wogau, Konrad. *Vier Studien zu Kants Kritik der Urteilskraft*. Uppsala: Uppsala Universitets Årsskrift, 1938.

Marino, Luigi. "Soemmering, Kant and the Organ of the Soul." In *Romanticism in Science: Science in Europe, 1790–1840*. Ed. Stefano Poggi and Maurizio Bossi, 47–74. Dordrecht: Kluwer, 1994.

May, James A. *Kant's Concept of Geography and Its Relation to Recent Geographical Thought*. Toronto: University of Toronto Press, 1970.

McLaughlin, Peter. *Kant's Critique of Teleology in Biological Explanation: Antinomy and Teleology*. Lewiston: E. Mellen Press, 1990.

Menzer, Paul. *Kants Lehre von der Entwicklung in Natur und Geschichte*. Berlin: Reimer, 1911.

Mischel, Theodore. "Kant and the Possibility of a Science of Psychology." In *Kant Studies Today*. Ed. Lewis W. Beck. Lasalle, IL: Open Court, 1969.

Model, Anselm. *Metaphysik und reflektierende Urteilskraft, Untersuchungen zur Transformierung des leibnizschen Monadenbegriffs in der KU*. Frankfurt: Athenaüm, 1987.

Paton, Herbert J. "Kant and the Errors of Leibniz." In *Kant Studies Today*. Ed. Lewis W. Beck, 301–21. Lasalle, IL: Open Court, 1969.

Piché, Claude. "Kant et les organismes non vivant." In *La nature*. Ed. L. Cournaraie and P. Dupond, 83–93. Paris: Ellipses, 2001.

Puech, Michel. *Kant et la causalité*. Paris: Vrin, 190.

Richards, Robert J. "Kant and Blumenbach on the *Bildungstrieb*: A Historical Misunderstanding." *Studies in History and Philosophy of Biological and Biomedical Sciences* 31, no. 1 (2000): 11–32.

Riese, Wolfgang. "Sur la théorie de l'organisme dans l'*Opus postumum* de Kant." *Revue philosophique* 3 (1965): 326–33.

Schlanger, Judith. *Schelling et la réalité finie*. Paris: Puf, 1966.

Schrader, George. "Status of Teleological Judgment in the Critical Philosophy." *Kant-Studien* 45 (1953–54): 204–34.

Sloan, Phillip. "Preforming the Categories: Eighteenth-Century Generation Theory and the Biological Roots of Kant's A Priori." *Journal of the History of Philosophy* 40, no. 2 (2002): 229–53.

———. "Kant and the History of Nature. The ambiguous heritage of the Critical Philosophy for Natural History." *Studies in History and Philosophy of Biological and Biomedical Sciences* 37 (2006): 7–22.

Stark, Werner. "Immanuel Kants physische Geographie—eine Herausforderung?" Lecture delivered as "Honorarprofessor Antrittsvorlesung" on May 4, 2001, at the Philipps-Universität, Marburg, Germany.

Strawson, Peter F. *The Bounds of Sense: An Essay on Kant's* Critique of Pure Reason. London: Routledge, 1995.

Sturm, Thomas. "Kant on Empirical Psychology: How Not to Investigate the Human Mind." In *Kant and the Sciences.* Ed. Eric Watkins, 163–84. Oxford: Oxford University Press, 2001.

Tonnelli, Giorgio. "La nécessité des lois de la nature au 18ème siècle et chez Kant en 1762." *Revue d'histoire des sciences* 12 (1959): 225–41.

———. "Von den verschiedenen Bedeutungen des Wortes Zweckmässigkeit in der Kritik der Urteilskraft." *Kant-Studien* 49 (1957–58): 154–66.

Ungerer, Emil. *Die Teleologie Kants und ihre Bedeutung für die Logik der Biologie.* Berlin, 1922.

Vuillemin, Jules. *Physique et métaphysique kantiennes.* Paris: Puf, 1987.

Wahsner, Renate. "Mechanism—Technizism—Organism: Der epistemologische Status der Physik als Gegenstand von Kants *Kritik der Urteilskraft.*" In *Naturphilosophie in Deutschen Idealismus.* Ed. Karen Gloy and Paul Burger, 1–23. Stuttgart: Fromann-Holzboog, 1993.

Walsh, William H. *Kant's Criticism of Metaphysics.* Edinburg: Edinburg University Press, 1975.

Watkins, Eric. *Kant's Conception of Causality.* Cambridge: Cambridge University Press, 2005.

Wellek, René. *Immanuel Kant in England, 1793–1838.* Princeton: Princeton University Press, 1931.

Westphal, Kenneth R. *Kant's Transcendental Proof of Realism.* Oxford: Oxford University Press, 2005.

Wike, Victoria. *Kant's Antinomies of Reason: Their Origin and Their Resolution.* Washington: University Press of America, 1982.

Wubnig, Julius. "The Epigenesis of Pure Reason." *Kant-Studien* 60 (1968): 147–52.

Zammito, John H. *The Genesis of Kant's* Kritik der Urteilskraft. Chicago: University of Chicago Press, 1992.

———. *Kant, Herder and the Birth of Anthropology.* Chicago: University of Chicago Press, 2002.

Zöller, Günther. "Kant on the Generation of Metaphysical Knowledge." In *Kant: Analysen—Probleme—Kritik.* Ed. Hariolf Oberer and Gerhardt Seel, 71–90. Wurtzburg: Königshausen & Neumann, 1988.

———. "From Innate to A *Priori*: Kant's Radical Transformation of a Cartesian-Leibnizian Legacy." *Monist* 72 (1989): 222–35.

Zumbach, Clark. *The Transcendant Science: Kant's Conception of Biological Methodology.* Den Haagen: Martinus Nijhoff, 1984.

Studies in the History and Philosophy of Biology: General Cultural History

Appel, Toby. *The Cuvier-Geoffroy Debate: French Biology in the Decades before Darwin*. Oxford: Oxford University Press, 1987.

Ashton, Rosemary. *The German Idea: Four English Writers and the Reception of German Thought 1800–1860*. Cambridge: Cambridge University Press, 1980.

Aulie, Richard. "Caspar Friedrich Wolff and His 'Theoria Generationis,' 1759." *Journal of the History of Medicine* 16 (1961): 124–44.

Bach, Thomas. *Biologie und Philosophie bei C. F. Kielmeyer und F. W. J. Schelling*, Schellingiana 12. Stuttgart: Bad-Cannstatt, 2001.

———."Kielmayer als 'Vater der Naturphilosophie'? Anmerkungen zu seiner Rezeption im deutschen Idealismus." *Philosophie des Organischen in der Goethezeit: Studien zur Werk und Wirkung des Naturforschers Carl Friedrich Kielmayer (1765–1844)*. Ed. Kai Torsten Kanz, 232–51. Boethius. Texte und Abhandlungen zur Geschichte der Mathematik und der Naturwissenschaft, 34. Stuttgart: Steiner, 1994.

Balan, Bernard. *L'ordre et le temps*. Paris: Vrin, 1979.

Baron, Wolfgang. "Die Anschauungen Johann Friedrich Blumenbachs über die Geschichtlichkeit der Nature." *Sudhoffs Archiv für Geschichte der Medizin und der Naturwissenschaften* 47 (1963): 19–26.

Barsanti, Giulio. "Buffon et l'image de la nature: de l'échelle des êtres à la carte géographique et à l'arbre généalogique." In *Buffon 88*. Ed. Jean Gayon, 255–95. Paris: Vrin, 1992.

———. "Lamarck and the Birth of Biology, 1740–1810." In *Romanticism in Science: Science in Europe, 1790–1840*. Ed. Stefano Poggi and Maurizio Bossi, 47–74. Dordrecht: Kluwer, 1994.

———. "La naissance de la biologie. Observations, théories, métaphysiques en France, 1740–1810." In *Nature, Histoire, Société, Mélanges offerts à Jacques Roger*. Ed. Roseline Rey, Claude Blanckaert, and Jean-Louis Fischer, 196–228. Paris: Klincksieck, 1995.

Bernasconi, Robert, ed. *Concepts of Race in the Eighteenth Century*. London: Thoemmes Continuum, 2001.

Bodemer, Charles W. "Regeneration and the Decline of Preformationnism in Eighteenth Century Embryology." *Bulletin of the History of Medicine* 38 (1962): 20–31.

Bowler, Peter J. *Evolution: The History of an Idea*. Berkeley: University of California Press, 1984.

———. "Preformation and Pre-existence in the Seventeenth Century: A Brief Analysis." *Journal of the History of Biology* 4, no. 2 (1971): 221–44.

Brady, Ron. "Form and Cause in Goethe's Morphology." In *Goethe and the Sciences: A Reappraisal*. Boston Studies in Philosophy of Science. Ed. Frederick Amrine, Harvey Zucker, and Francis J. Wheeler, ch. 5. Dordrecht: Reidel, 1987.

Breidbach, Olaf, and Michael Ghiselin. "Lorenz Oken and Naturphilosophie in Jena, Paris, and London." *History and Philosophy of the Life Sciences* 24 (2002): 219–47.

Breidbach, Olaf, Hans-Joachim Fliedner, and Klaus Ries, eds. *Lorenz Oken (1779–1851). Ein politischer Naturphilosoph*. Weimar: Hermann Böhlaus, 2001.

Breidbach, Olaf. "Die Geburt des Lebendigen—Embryogenese der Formen oder Embryologie der Natur?—Anmerkungen zum Bezug von Embryologie und Organismustheorien vor 1800." *Biologisches Zentralblatt* 114 (1964): 191–99.

Brown, Theodore M. "Descartes, Dualism and Psychosomatic Medicine." In *Anatomy of Madness: Essays in the History of Psychiatry*. Ed. Roy Porter, William Bynum, and Michael Shepherd, 41–63. 3 vols. London: Tavistock Press, 1985.

———. "From Mechanism to Vitalism in Eighteenth-Century English Physiology." *Journal of the History of Biology* 7, no. 2 (1974): 179–216.

Bynum, William F. "Nosology." In *Companion Encyclopedia of the History of Medicine*. Ed. Roy Porter and William F. Bynum, 335–56. London: Routledge, 1993.

Canguilhem, Georges. *La formation du concept de réflexe aux XVIIᵉ et XVIIIᵉ siècle*. Paris: Puf, 1955.

Canguilhem, George, George Lapassade, George Piquemal, and Jacques Ullman. *Du développement à l'évolution au 19ème siècle*. Paris: Puf, 1962.

Carlyon, Clement. *Early Years and Late Reflections*. London: Whittaker, 1836.

Coleman, William. *Biology in the 19th Century: Problems of Form, Function and Transformation*. Cambridge: Cambridge University Press, 1979.

———. "Limits of the Recapitulation Theory: Carl Friedrich Kielmayer's Critique of the Presumed Parallelism of Earth History, Ontogeny, and the Present Order of Organisms." *Isis* (1973): 341–50.

Corsi, Pietro. *The Age of Lamarck: Evolutionary Theories in France, 1790–1830*. Berkeley: University of California Press, 1988.

Cunningham, Andrew. "The Pen and the Sword: Recovering the Disciplinary Identity of Physiology and Anatomy before 1800–II: Old Physiology—the Pen." *Studies in History and Philosophy of Biological and Biomedical Sciences* 34, no. 1 (2003): 51–76.

Dawson, V. "Regeneration, Parthenogenesis, and the Immutable Order of Nature." *Archives of Natural History* 18 (1991): 309–21.

De Diesbach, Ghislain. *Madame de Staël*. Paris: Perrin, 1997.

Duchesneau, François. "Epigénèse et évolution: prémisses historiques." *Annales d'Histoire et de Philosophie du vivant* 6 (2002): 177–203.

———. "Haller et les théories de Buffon et C. F. Wolff sur l'épigenèse." *History and Philosophy of the Life Sciences* 1 (1985): 65–100.

———. *La Physiologie des Lumières. Empirisme, modèles et théories*. Den Haagen: Martinus Nijhoff, 1982.

———. "Vitalism in Late 18th Century Physiology: The Cases of Barthez, Blumenbach, and John Hunter." In *William Hunter and the 18th-Century Medical World*. Ed. William F. Bynum and Roy Porter, 259–95. Cambridge: Cambridge University Press, 1985.

Duchet, Michèle. *Anthropologie et histoire au siècle des Lumières*. Paris: Albin Michel, 1995.

Ecker, Alexander. *Lorenz Oken. Eine biographische Skizze. Gedächtnisrede zu dessen hundertjährigen Geburtstagsfeier gesprochen in der zweiten öffentlichen Sitzung der 52. Versammlung deutscher Naturforscher und Aerzte zu Baden-Baden am 20. September 1879 von Alexander Ecker. Durch erläuternde Zusätze und Mitteilungen aus Oken Briefwechsel vermehrt. Mit dem Portrait Oken's und einem Facsimile der Nr. 195 des*

I. Bandes der Isis. Stuttgart: E. Schweizerbart, 1880. Translated by Alfred Tulk, *Lorenz Oken. A biographical sketch.* London: Kegan Paul, 1883.

Fabbri Bertoletti, Stefano. "The Anthropological Theory of Johann Friedrich Blumenbach." In *Romanticism in Science: Science in Europe, 1790–1840.* Ed. Stefano Poggi and Maurizio Bossi, 103–25. Dordrecht: Kluwer, 1994.

Farber, Paul L. "Buffon and the Concept of Species." *Journal of the History of Biology* 5 (1972): 259–84.

Figlio, Karl. "Theories of Perception and the Physiology of Mind in the Late Eighteenth Century." *History of Science* 12 (1975): 177–212.

Gaissinovich, A. E. "Le rôle du Newtonianisme dans la renaissance des idées épigénetiques en embryologie du XVIIIe siècle." In *Actes du XIe Congrès International d'Histoire des Sciences* 5 (1968): 105–10.

Gasking, Elizabeth. *Investigations into Generation 1651–1828.* Baltimore: John Hopkins University Press, 1967.

Glass, Bentley, Oswei Temkin, and William Strauss, eds. *Forerunners of Darwin, 1745–1859.* Baltimore: Johns Hopkins University Press, 1959.

Gould, Stephen Jay. *Ontogeny and Phylogeny.* Cambridge: Belknap Press, 1977.

Gregory, Frederick. "Kant's Influence on Natural Scientists in the German Romantic Period." In *New Trends in the History of Science.* Ed. Robert Paul, W. Visser, H. Bos, L. Palm, and H. Snelders, 53–72. Amsterdam: Rodopi, 1989.

———. "'Nature is an organized whole': Fries's Reformulation of Kant's Philosophy of Organism." In *Romanticism in Science: Science in Europe, 1790–1840.* Ed. Stefano Poggi and Maurizio Bossi, 91–101. Dordrecht: Kluwer, 1994.

Haigh, Elizabeth. "Vitalism, the Soul, and Sensibility: The Physiology of Théophile Bordeu." *Journal of the History of Medicine* 31 (1976): 30–41.

Harman, Peter. *Metaphysics and Natural Philosophy: The Problem of Substance in Classical Physics.* New Jersey: Barnes and Noble, 1982.

Hatfield, Gary. "Empirical, Rational and Transcendental Philosophy: Psychology as Science and as Philosophy." In *Cambridge Companion to Kant.* Ed. Paul Guyer, 200–228. Cambridge: Cambridge University Press, 1992.

———. "Remaking the Science of Mind: Psychology as Natural Science." In *Inventing Human Sciences.* Ed. Christopher Fox, Robert Wokler, and Roy Porter, 184–232. Berkeley: University of California Press, 1995.

Heimann, P. M. "'Nature is a perpetual worker': Newton's Aether and Eighteenth-Century Natural Philosophy." *Ambix* 20 (1973): 1–25.

———. "Voluntarism and Immanence: Conceptions of Nature in Eighteenth-Century Thought." *Journal of the History of Ideas* 39 (1978): 271–83.

Heimann, P. M., and J. E. McGuire. "Newtonian Forces and Lockean Powers: Concepts of Matter in Eighteenth-Century Thought." *Historical Studies in the Physical Sciences* 3 (1971): 233–306.

Hintzsche, Erich. "Einige kritische Bemerkungen zur Bio- und Ergographie Albrecht von Hallers." *Gesnerus* 16 (1959): 1–15.

Hoffheimer, Michael H. "Maupertuis and the Eighteenth-Century Critique of Preexistence." *Journal of the History of Biology* 15, no. 1 (1982): 119–44.

Hoquet, Thierry. "La comparaison des espèces. Ordre et méthode dans l'*Histoire naturelle* de Buffon." *Corpus. Revue de Philosophie* 43 (2003): 355–416.

Jackson, Heather J. "Coleridge's Collaborator, Joseph Henry Green." *Studies in Romanticism* 21 (1982): 161–79.

Jackson, Stanley. "Force and Kindred Notions in 18th Century Neurophysiology and Medical Psychology." *Bulletin of the History of Medicine* 44 (1970): 539–54.

Jacob, Margaret. *The Radical Enlightenment: Pantheists, Freemasons and Republicans.* London: Allen and Unwin, 1981.

Jacyna, L. Stephen. "Immanence or Transcendence: Theories of Life and Organization in Britain, 1790–1835." *Isis* 74 (1983): 311–29.

———. "The Romantic Program and the Reception of Cell Theory in Britain." *Journal of the History of Biology* 17, no. 1 (1984): 13–48.

———. "Romantic Thought and the Origins of Cell Theory." In *Romanticism and the Sciences*. Ed. Andrew Cunningham and Nick Jardine, 161–68. Cambridge: Cambridge University Press, 1990.

von Jäger, Georg-Friedrich. "Ehrengedächtniss des Königl. Württembergischem Staatsraths von Kielmeyer." *Nova Acta Academiae Caesarae Leopoldina-Carolinae Germanicae Naturae Curiosorum* 21, no. 2 (1845): i–xcii.

Jahn, Ilse. "Georg Forsters Lehrkonzeption für eine 'Allgemeine Naturgeschichte' (1786–1793) und seine Auseinandersetzung mit Caspar Friedrich Wolffs 'Epigenesis'-Theorie." *Biologisches Zentralblatt*, 114 (1995): 200–206.

———. "On the Origin of Romantic Biology and Its Further Development at the University of Jena between 1790 and 1850." In *Romanticism in Science: Science in Europe, 1790–1840*. Ed. Stefano Poggi and Maurizio Bossi, 75–89. Dordrecht: Kluwer, 1994.

Jardine, Nicholas. "*Naturphilosophie* and the Kingdoms of Nature." In *Cultures of Natural History*. Ed. Nicholas Jardine, John Secord, and E. Sparry, 230–45. Cambridge: Cambridge University Press, 1996.

———. *Scenes of Inquiry: On the Reality of Questions in the Sciences.* Oxford: Clarendon Press, 1991.

Kanz, Kai Torsten, ed. *Kielmeyer-Bibliographie: Verzeichnis der Litteratur von und über den Naturforscher Carl Friedrich Kielmeyer (1765–1844)*, Quellen der Wissenschaftsgeschichte, 1. Stuttgart: Verlag für Geschichte der Naturwissenschaft und der Technik, 1991.

———. *Philosophie des Organischen in der Goethezeit*. Boethius, Texte und Abhandlungen zur Geschichte der Mathematik und der Naturwissenschaft, 34. Stuttgart, 1994.

Klein, Marc. *Regards d'un biologiste*. Paris: Hermann, 1980.

Larson, James L. *Interpreting Nature: The Science of Living Form from Linneaus to Kant.* Baltimore: Johns Hopkins University Press, 1994.

———. *Reason and Experience: The Representation of Natural Order in the Work of Carl von Linné*. Berkeley: University of California Press, 1971.

———. "Vital Forces: Regulative Principles or Constitutive Agents? A Strategy in German Physiology, 1786–1802." *Isis* 70 (1979): 235–49.

Levere, Trevor. *Poetry Realized in Nature: Samuel Taylor Coleridge and Early Nineteenth-Century Science*. Cambridge: Cambridge University Press, 1981.

McLaughlin, Peter."Blumenbach und der Bildungstrieb: zum Verhältnis von epigenetischer Embryologie und typologischem Artbegriff." *Medizinhistorisches Journal* 17 (1982): 357–72.

————. "Soemmering und Kant: Uber das Organ der Seele und den Streit der Fakultäten." *Soemmering-Forschungen* 1 (1985): 188–95.

————. *What Functions Explain: Functional Explanation and Self-Reproducing Systems.* Cambridge: Cambridge University Press, 2001.

Lenoir, Timothy. "The Eternal Laws of Form: Morphotypes and the Conditions of Existence in Goethe's Biological Thought." In *Goethe and the Sciences: A Reappraisal.* Boston Studies in Philosophy of Science. Ed. Frederick Amrine, Harvey Zucker, and Francis J. Wheeler, 17–28. Dordrecht: Reidel, 1987.

————. "The Göttingen School and the Development of Transcendantal *Naturphilosophie* in the Romantic Era." *Studies in the History of Biology* 5 (1981): 111–205.

————. "Generational Factors in the Origin of *Romantische Naturphilosophie.*" *Journal of the History of Biology* 11, no. 1 (1978): 57–100.

————. "Kant, Blumenbach and Vital Materialism in German Biology." *Isis* 71 (1980): 77–108.

————. "Morphotypes in Romantic Biology." In *Romanticism and the Sciences.* Ed. Andrew Cunningham and Nick Jardine, 120–29. Cambridge: Cambridge University Press, 1990.

————. *The Strategy of Life: Teleology and Mechanism in Nineteenth-Century German Biology.* Dordrecht: Reidel, 1982.

Lepenies, Wolff. *Das Ende der Naturgeschichte: Wandel kulturellen Selbstverständlichkeiten in des Wissenschaften des 18. und 19. Jahrhundert.* Munich: Hansen, 1976.

Lieber, Hans-Joachim. "Kants Philosophie des Organischen und die Biologie seiner Zeit." *Philosophia naturalis* 1 (1950): 553–70.

Liedman, S. E. *Det organiska livet i tysk debatt 1795–1845.* Lund: Berlinska, 1966.

Linden, Mareta. *Untersuchungen zum Anthropologiebegriff der 18. Jahrhundert.* Bern: Herbert Lang, 1979.

Lopez-Beltràm, Carlos. "Natural Things and Non-natural Things: The Boundaries of Heredity in the 18th Century." *A Cultural History of Heredity: 17th and 18th Centuries.* Preprint 222. Berlin: Max Planck Institut, 2002.

Mendelsohn, Everett. "Physical Models and Physiological Concepts: Explanation in Nineteenth Century Biology." In *Boston Studies in the Philosophy of Science.* Ed. Max Cohen and Marx Wartofsky, 127–50. New York: Humanities Press, 1965.

Mocek, Reinhard. "Caspar Friedrich Wolffs Epigenesis-Konzept—ein Problem im Wandel der Zeit." *Biologisches Zentralblatt* 114 (1995): 179–90.

Müller-Sievers, Helmut. *Epigenesis: Naturphilosophie im Sprachdenken Wilhelm von Humboldts.* Paderborn: Schöningh, 1993.

————. *Self-Generation: Biology, Philosophy and Literature around 1800.* Stanford: Stanford University Press, 1997.

Oppenheimer, Jane. *Essays in History of Embryology.* Cambridge, MA: MIT Press, 1967.

Osler, Margaret. "John Locke and the Changing Ideal of Scientific Knowledge." *Journal of the History of Ideas* 31 (1970): 3–16.

Ospovat, Dov. *The Development of Darwin's Theory: Natural History, Natural Theology, and Natural Selection, 1838–1859.* Cambridge: Cambridge University Press, 1981.

Porter, Roy. "Medical Science and Human Science." In *Inventing Human Sciences.* Ed. Christopher Fox, Robert Wokler, and Roy Porter, 53–87. Berkeley: University of California Press, 1995.

Pross, Wolfgang. "Herder und die Anthropologie seiner Zeit." In *Herder und die Anthropologie der Aufklärung*. Ed. Wolfgang Proß, 1128–216. Darmstadt: Wissenschaftliche Buchgesellschaft, 1987.

———. "Herders und Kielmeyers Begriff der organischen Kräfte." In *Philosophie des Organismen in der Goethezeit: Studien zu Werk und Wirkung des Naturforschers Carl Friedrich Kielmeyer (1765–1844)*. Ed. Kai Torsten Kanz, 81–99. Boethius: Texte und Abhandlungen zur Geschichte der Mathematik und der Naturwissenschaft, 34. Stuttgart: Steiner, 1994.

Ratcliff, Marc. "Abraham Trembley's Strategy of Generosity and the Scope of Celebrity in the Mid-eighteenth Century." *Isis* 95 (2004): 555–75.

Ratcliffe, Matthew. "The Function of Functions." *Studies in History and Philosophy of Biological and Biomedical Sciences* 31, no. 1 (2000): 113–33.

Rehbock, Philip H. *The Philosophical Naturalists: Themes in Early Nineteenth-Century British Biology*. Madison: University of Wisconsin Press, 1983.

———. "Transcendental Anatomy." In *Romanticism and the Sciences*. Ed. Andrew Cunningham and Nick Jardine, 144–60. Cambridge: Cambridge University Press, 1990.

Reill, Peter Hans. "Analogy, Comparison and Active Living Forces: Late Enlightenment Responses to the Critiques of Causal Analysis." In *The Sceptical Tradition around 1800*. Ed. Karl van der Zande and Richard Popkins, 203–11. Dordrecht: Kluwer, 1998.

———. "Anthropology, Nature and History in the Late Enlightenment: The Case of Friedrich Schiller." In *Schiller als Historiker*. Ed. Otto Dann, Norbert Oellers, and Ernst Osterkamp, 243–65. Stuttgart: Metzler, 1995.

———. "Anti-mechanism, Vitalism and Their Political Implications in Late Enlightened Scientific Thought." *Francia*, Band 16/2, 195–212. Sigmarigen: Jan Thorbeck Verlag, 1990.

———. "Between Mechanism and Hermeticism: Nature and Science in the Late Enlightenment." In *Frühe Neuzeit-Frühe Moderne?* Ed. Rudolf Vierhaus, 393–421. Göttingen: Vandenhoeck and Ruprecht, 1992.

———. "Between Preformation and Epigenesis: Kant, Physicotheology and Natural History." *New Essays on the Precritical Kant*. Ed. Tom Rockwell, 161–81. New York: Humanity Books, 2001.

———. "Buffon and Historical Thought in Germany and Great Britain." In *Buffon 88*. Ed. Jean Gayon, 662–90. Paris: Vrin, 1992.

———. "Science and the Construction of the Cultural Sciences in Late Enlightenment Germany: The Case of Wilhelm von Humboldt." *History and Theory* 33, no. 3 (1994): 345–66.

———. "Science and the Science of History in the *Spätaufklärung*." In *Aufklärung und Geschichte*. Ed. Hermann E. Boedeker, Georg Iggers, James Knudsen, and Peter Hans Reill. Göttingen: Vandenhoeck and Ruprecht, 1986.

———. *Vitalizing Nature in the Enlightenment*. Berkeley: University of California Press, 2005.

———. "Vitalizing Nature and Naturalizing the Humanities in the Late Eighteenth Century." *Studies in Eighteenth Century Culture*, vol. 28. Ed. J. C. Hayes and T. Erwin, 361–81. Baltimore: John Hopkins University Press, 1999.

Rheinberger, Hans-Jorg. "Aspekte des Bedeutungswandels im Begriff organismischer Ähnlichkeit vom 18. zum 19. Jahrhundert." *History and Philosophy of the Life Sciences* 8 (1986): 237–50.

———. "Über Formen und Gründe der Historisierung biologischer Modelle von Ordnung und Organisation am Ausgang des 18. Jahrhunderts." In *Gesellschaftliche Bewegung und Naturprozeß*. Ed. Manfred Hahn and Hans-Jörg Sandkühler, 71–81. Cologne: Paul-Rugenstein, 1981.

Richards, Eveleen. "'Metaphorical Mystifications': The Romantic Gestation of Nature in British Biology." In *Romanticism and the Sciences*. Ed. Andrew Cunningham and Nick Jardine, 130–43. Cambridge: Cambridge University Press, 1990.

Richards, Robert J. *The Meaning of Evolution: The Morphological Construction and Ideological Reconstruction of Darwin's Theory*. Chicago: Chicago University Press, 1992.

———. *The Romantic Conception of Life: Science and Philosophy in the Age of Goethe*. Chicago: Chicago University Press, 2002.

Riedel, Wolfgang. "Influxus physicus und Seelenstärke." In *Anthropologie und Literatur um 1800*. Ed. Jürgen Barkhoff and Eda Sagarra, 24–52. Munich: Iudicium, 1992.

Rittersbuch, Philip C. *Overtures to Biology: The Speculations of Eighteenth-Century Naturalists*. New Haven: Yale University Press, 1964.

Roe, Shirley. *Matter, Life and Generation: Eighteenth-Century Embryology and the Haller-Wolff Debate*. Cambridge: Cambridge University Press, 1980.

———. "Rationalism and Embryology: Caspar Friedrich Wolff's Theory of Epigenesis." *Journal of the History of Biology* 12, no. 1 (1979): 1–43.

Roger, Jacques. "Buffon et l'introduction de l'histoire dans l'histoire naturelle." In *Buffon 88*. Ed. Jean Gayon, 193–205. Paris: Vrin, 1992.

———. *Buffon. Un philosophe au jardin du Roi*. Paris: Fayard, 1989.

———. *Les sciences de la vie dans la pensée française au 18ème siècle*. 1963; 2nd ed., Paris: Albin Michel, 1993.

Ruse, Michael. *The Darwinian Revolution*. Chicago: University of Chicago Press, 1979.

Sabra, Abdelhamid I. "The Appropriation and Subsequent Naturalization of Greek Science in Medieval Islam." *History of Science* 25 (1987): 223–43.

Schmitt, Stéphane. *Histoire d'une question anatomique: la répétition des parties*. Paris: Editions du Muséum d'Histoire Naturelle, 2006.

———. *Les textes embryologiques de Christian Heinrich Pander*. Paris: Brepols, 2003.

———. "Type et métamorphose dans la morphologie de Goethe, entre Classicisme et Romantisme." *Revue d'Histoire des Sciences* 54 (2001): 495–522.

Shell, Susan Meld. *The Embodiment of Reason: Kant on Spirit, Generation and Community*. Chicago: University of Chicago Press, 1996.

Sloan, Phillip R. "Buffon, German Biology, and the Historical Interpretation of Biological Species." *British Journal for the History of Science* 12 (1979): 109–53.

———. "The Gaze of Natural History." In *Inventing Human Sciences*. Ed. Christopher Fox, Robert Wokler, and Roy Porter, 112–53. Berkeley: University of California Press, 1995.

———. "Le Muséum de Paris vient à Londres." In *Le Muséum au premier Siècle de son Histoir*. Ed. Claude Blanckaert, Claudine Cohen, Pietro Corsi, and Jean-Louis Fischer, 606–47. Paris: Editions du Muséum, 1997.

———. "Natural History, 1670–1802." In *Companion to the History of Modern Science*. Ed. R. C. Olby, G. N. Cantor, J. R. R. Christie, and M. J. S. Hodge, 295–313. London: Routledge, 1996.

———. "Organic Molecules Revisited." In *Buffon 88*. Ed. Jean Gayon, 415–38. Paris: Vrin, 1992.

———. "The Question of Natural Purpose." In *Evolution and Creation*. Ed. Ernan MacMullin, 121–50. Notre Dame: University of Notre Dame Press, 1985.

———. "Whewell's Philosophy of Discovery and the Archetype of the Vertebrate Skeleton: The Role of German Philosophy of Science in Richard Owen's Biology." *Annals of Science* 60 (2003): 39–61.

Sloan, Phillip R., and John Lyon. *From Natural History to the History of Nature*. Notre Dame: University of Notre Dame Press, 1981.

Smith, Robert. "The Background of Physiological Psychology in Natural Philosophy." *History of Science* 11 (1973): 75–123.

Srohl, Jean. *Lorenz Oken und Georg Büchner*. Zurich: Verlag der Corona, 1936.

Steigerwald, Joan. "Instruments of Judgement: Inscribing Processes in Late 18th Century Germany." *Studies in History and Philosophy of Biology and Biomedical Sciences* 33 (2002): 79–131.

Temkin, Oswei. "Basic Science, Medicine and the Romantic Era." *Bulletin of the History of Medicine* 37, no. 2 (1963): 97–129.

———. "German Concepts of Ontogeny and History around 1800." In *The Double Face of Janus and Other Essays*, 273–389. Baltimore: John Hopkins University Press, 1977.

———. "Materialism in French and German Physiology of the Early Nineteenth Century." *Bulletin of the History of Medicine* 20 (1946): 322–30.

Terrall, Mary. *The Man Who Flattened the Earth*. Chicago: University of Chicago Press, 2002.

Thackray, Arnold. *Atoms and Powers: An Essay on Newtonian Matter-Theory and the Development of Chemistry*. Cambridge, MA: Harvard University Press, 1970.

Wokler, Robert. "Anthropology and Conjectural History in the Enlightenment." In *Inventing Human Sciences*. Ed. Christopher Fox, Robert Wokler, and Roy Porter, 31–52. Berkeley: University of California Press, 1995.

Wood, Paul B. "The Science of Man." In *Cultures of Natural History*. Ed. Nicholas Jardine, John Secord, and Emma Sparry, 197–210. Cambridge: Cambridge University Press, 1996.

Yolton, John. *Locke and French Materialism*. Oxford: Clarendon Press, 1991.

———. *Thinking Matter: Materialism in Eighteenth-Century Britain*. Minneapolis: University of Minnesota Press, 1983.

Zammito, John. "Epigenesis: Concept and Metaphor in Herder's *Ideen*." In *Vom Selbstdenken: Aufklärung und Aufklärungskritik in Johann Gottfried Herders "Ideen zur Philosophie der Geschichte der Menschheit."* Ed. Rudolf Otto and John Zammito, 131–45. Heidelberg: Synchron, 2001.

———. "'Method' versus 'Manner'? Kant's Critique of Herder's *Ideen* in the Light of the Epoch of Science, 1790–1820." In *Herder-Jahrbuch/Herder Yearbook 1998*. Ed. Hans Adler and Wulf Koepke, 1–26. Stuttgart: Metzler, 1998.

Zaunick, Rudolf, ed. *Aus Leben und Werk von Lorenz Oken, dem Begründer der deutschen Naturforscherversammlungen*. Friburg in Brisgau, 1938.

CONTRIBUTORS

JEAN-CLAUDE DUPONT. Professor at the Université d'Amiens and researcher at the Institut d'Histoire et de Philosophie des Sciences et des Techniques (CNRS, Paris). He is the author of *Histoire de la neurotransmission* (Paris: PUF, "Histoire, science et société," 1999), and of a critical anthology of embryology entitled *Du feuillet au gène. Une histoire de l'embryologie moderne (fin 18ème-XXème siècle)* (Paris: éd. rue d'Ulm, 2004, with Stéphane Schmitt). He edited the French translation of Wolff's *De formatione intestinorum* (Paris: Brepols, 2003), has published numerous papers on history and epistemology of physiology, neurosciences, and embryology, and is editing a collective volume on the history of memory.

MARK FISHER. Lecturer at Penn State University, University Park. He will receive his Ph.D in Philosophy from Emory University through defense of his dissertation "Organisms and Teleology in Kant's Natural Philosophy."

PHILIPPE HUNEMAN. Researcher at the Institut d'Histoire et de Philosophie des Sciences et des Techniques (CNRS-Université Paris I Sorbonne, Paris). Author of *Bichat. La vie et la mort* (Paris: PUF, 1998) and *Métaphysique et biologie. Kant et la constitution du concept d'organisme* (Paris: PUF, forthcoming), he has published papers on classical metaphysics and Kant and about the rise of psychiatry in the nineteenth century; he is currently working on the philosophy of evolutionary biology.

ROBERT J. RICHARDS. Professor at the University of Chicago, and Director of the Fishbein Center for the History of Biology. Author of *Darwin and the Emergence of Evolutionary Theories of Mind and Behavior* (Chicago: University of Chicago Press, 1986), *The Meaning of Evolution* (Chicago: University of Chicago Press, 1994), and *The Romantic Conception of Life* (Chicago: University of Chicago Press, 2002). He has published numerous papers on the history of biology in the nineteenth century, the history of psychology, and the philosophy of biology. He is currently publishing a book on Haeckel and evolutionism (*The Tragic Sense of Life: Ernest Haeckel and the Struggle over Evolutionary Thought*, Chicago: University of Chicago Press).

STÉPHANE SCHMITT. Researcher at the REHSEIS center (CNRS, Paris—Université Paris VII). Author of *Du feuillet au gène. Une histoire de l'embryologie moderne (fin 18ème-XXème siècle)* (Paris: éd. rue d'Ulm, 2004), with Jean-Claude Dupont, and of a history of the concept of homology (*Histoire d'une question anatomique: la répétition des parties* (Paris: Editions du

Museum d'Histoire Naturelle, 2005). He edited and translated into French the embryological works of Christian Heinrich Pander (Paris: Brepols, 2003), and edited works of Link, Kielmayer, and Oken (Paris: Brepols, 2006). Author of papers on German and French comparative anatomy, Goethe, *Naturphilosophie*, and neo-Lamarckism, he is currently working on editing and introducing Herder's work in the life sciences.

PHILLIP R. SLOAN. Professor at the University of Notre-Dame (Indiana), and in the Program of Liberal Studies/Program in History and Philosophy of Science. Author of *From Natural History to the History of Nature* (1981), he is also the editor of Owen's lectures in comparative anatomy and of the collected volume *Controlling Our Destiny* (Notre-Dame: University of Notre Dame Press, 2000), about the Human Genome Project. He has published many papers on nineteenth-century German biology, late nineteenth-century biology in Britain and the emergence of evolutionism, natural history in the 1700s, and teleology.

JOHN H. ZAMMITO. Professor at Rice University (Houston, Texas). He is the author of *The Genesis of Kant's Critique of Judgment* (Chicago: University of Chicago Press, 1993), *Kant, Herder and the Birth of Anthropology* (Chicago: University of Chicago Press, 2001), and *A Nice Derangement of Epistemes: Post-positivism in the Study of Science from Quine to Latour* (Chicago: University of Chicago Press, 2004). He has published numerous papers on the history of Kantian philosophy and on the history of human sciences. He is currently working on the epistemology of intellectual history.

Index

adaptation, 5 35, 57, 82
agent, 46, 158, 162
Allison, Henry, 24n18, 28n49, 29n59, 95n24
analogy, 5, 28n50, 45, 51, 54, 56–58, 60–64, 78–79, 93n10, 98n52, 9n62, 119n22, 157, 162
animal soul, 53, 67
Anlagen (dispositions), 5, 57, 58, 63, 73, 77, 82
antinomy, 10, 28n53, 29n60, 74, 81, 95n24, 98n50

Beiser, Friedrick, 24, 31n97, 35n104, 99n65, 133n8
Blumenbach, Johann Friedrich, 1–3, 8–9, 12–13, 15–19, 25n34, 26n42, 31n71, 43–47, 49n28, 51, 54, 62, 65, 69n25, 73n68, 82, 85–88, 92, 96n29, 97n48, 99n56, 123, 125–28, 133n5, 153, 158–59, 163–64
body, 23n17, 37, 42 49 52, 53, 59, 71, 77, 82–83, 86, 90, 106, 107, 109, 131, 159, 164, 167
Boissier de Sauvages, François, 7
Bonnet, Charles, 37, 42–45, 57, 62, 97, 108–12, 120n26
Brandt, Reinhardt, 29n61, 115n15
Buffon, George Louis Leclerc Comte de, 1, 5, 7–9, 25n33, 27n45, 33n93, 48, 54, 68n19, 78, 82–84, 91, 109, 113n5, 117n19, 135n28, 156

Caneva, Kenneth, 2, 12, 16, 30–31, 96, 133
causality, 5–6, 10–12, 24, 31–32, 64, 74, 84, 86, 94–95, 98–99, 103, 111, 146
cellular theory, 2, 47
classification, 101, 123
Coleridge, Samuel T., 149, 151, 153–55, 164–67, 169

contingency, 53, 75–80, 90–92, 102–3, 108, 110–11
creativity, 145, 151
Critique of pure reason, 4, 10, 13, 78–79, 82, 90, 93, 98, 143
Cuvier, Georges, 17–18, 33, 126, 129, 132, 159, 160–63

Darwin, Charles, 18, 33, 144
Descartes, René, 11, 44, 53, 55, 63, 69
design, 54, 65, 79, 88, 94n19, 95n22, 104, 109, 115n13, 146, 156
dispositions. *See Anlagen*

educt, 60, 64, 71n54
embryo, 6, 16, 40, 43, 47, 57
embryology, 2, 20–21, 38, 48, 54, 66, 81–87, 128, 149
emergence, 54–55, 62, 169
epigenesis, 11–15, 20–21, 42–45, 53–56, 60–66, 102–3, 108–12, 120n26
experimental physics, 52, 56, 74n87

Fichte, Johann, 14, 150, 152
force(s): attractive force, 46; *Lebenskraft,* 46, 124, 128; vegetative forces, 40, 46; vis essentialis, 39, 42–46, 54, 83–87, 97n48; vital forces, 12, 31n69, 81, 123–26, 132
Forster, Georg, 9, 58, 62, 65, 73n69, 157–58, 168n37
Friedman, Michael, 14, 29n54, 79n32, 66n6
function, 20, 67n7, 80, 95n22, 102, 107, 111, 128–32, 161, 164, 168 ; succession of functions, 131

generatio aequivoca, 64, 72n60
generation, theory of, 6, 21, 39, 47, 52, 57
Ginsborg, Hannah, 22n2, 24n27, 100n73

Girtanner, Friedrich, 7, 25n34, 26n42
Green, Joseph Henri, 155–64
Grier, Michelle, 28n49, 29n58, 94n16, 95n24, 98n50
Guyer, Paul, 5, 24n18, 25n44, 28n49, 32n80, 92n1

Haller, Albrecht von, 1, 12, 38–40, 44, 57–58, 69n27, 70n42, 83–84, 90, 97n35, 100, 108–9, 111, 113n5, 117n19, 120n26, 124–25, 132n3
Harvey, William, 11, 42–43, 64, 68n18, 113n1
Hatfield, Gary, 27n43
heart, 40–42, 44, 83, 97n35, 128–29, 131
Hegel, Georg Wilhelm Friedrich, 10, 17, 34n99, 150
Herder, Johann Gottfried, 8, 13, 19, 36n115, 54, 58–59, 62, 65, 89, 113n5, 126, 133n8, 142, 143–45, 158, 169n43
heredity, 4, 12, 30n65, 53, 78, 84, 91, 96n30, 108–9
Histoire naturelle, 25n32, 27n45
Hunter, John, 22, 156
hylozoism, 12, 21, 40n67, 52, 56–59, 61–63, 67n6

inscrutability, 59
intellectus archetypus, 137, 146, 192
intestine, 38, 43–44, 47
Invisibility, 37, 41, 43
irritability, 54, 84, 90, 124–28

Jardine, Nicholas, 16–18, 30n66, 32n89, 34n87, 35n105, 96n34, 127, 134n15

Kant's essays on races *(Verschiedenen Rassen.),* 80, 96, 115n14, 168n36
Keime, 5, 13, 58–59, 62–65, 82, 96n28
Kielmeyer, Carl Friedrich, 123, 126–28, 133n11
Kitcher, Philip, 5, 24n22, 28n52, 94n17

Lamarck, Jean Baptiste, 84, 97, 149–50, 159–61, 164, 169n49
Larson, James, 2, 16, 32n88, 117n19, 120n26, 123n1, 150, 165n2

lawlikeness, 9, 14, 27n48, 29n54, 78, 90–92, 121
layers, 43, 47, 83–84, 121
Lebrun, Gérard, 5, 11, 24n19, 28n52, 29n62, 92n1, 94n17
Leibniz, Gottfried, 1, 11, 13, 25n43, 33n91, 38n57, 61, 75–76, 85, 98n50, 120n23
Lenoir, Timothy, 2, 3, 12, 15, 17–20, 25n34, 30n66, 32n87, 33n95, 69n22, 96n30, 97n43, 98n48, 123, 127, 128, 132, 133n35, 134n20, 150
Linnaeus, 5–7, 68, 124, 157
Link, Heinrich Friedrich, 16, 128, 134n21
Locke, John, 67n11, 157
Longuenesse, Béatrice, 28n53
Löw, Reinhardt, 23n7, 69n32

machine, 24n27, 45, 66, 85, 98n50, 107
materialism, vital, 53, 64, 67n13
matter: inert, 52, 55, 65, 150; thinking, 53
Maupertuis, Pierre-Louis, 1, 48, 54, 57, 78, 82–84, 93n13, 108–9, 113n1, 121n30
Mc Laughlin, Peter, 5, 10, 20, 23n4, 24n27, 28n53, 29n55, 69n24, 73n71, 74n86, 95n24, 132n1, 133n5
mechanism, 6, 10, 19, 21, 29n56, 45–47, 53–54, 63–64, 85, 95n24, 101–2
membranes, 42, 44, 47
Metaphysical foundations of natural sciences, 56, 65, 69n31, 78, 113n2, 150
morphology, 3, 16–18, 22, 33n92, 34n98, 131–32, 144–45

natural history, 161, 4, 7–8, 21–22, 25n30, 31n71, 57, 92, 101, 103–6, 109–12, 129, 150, 153–57
Naturgeschichte, 105, 157, 172
Naturphilosophie, 8, 15, 17–21, 25n104, 34n99, 35n104, 129
Newtonianism, 60, 65, 150
Nutrition, 40, 45, 47, 49, 82, 126

Oken, Lorenz, 13, 17, 21, 34n97, 34n100, 123, 128–31, 134n22 order of nature, 4–5, 8, 28n53, 65, 78–80, 89–90, 93n16

Organism, 6–7, 15, 22, 54–55, 59, 85–88, 102–3, 125–27, 145–47

Organization, 7, 39, 49, 94, 98, 126; original organization, 2, 6–7, 16, 63–64, 85, 103

Ospovat, Dov, 18, 33n92, 35n109

Owen, Richard, 17–18, 22, 34n100, 149, 164–65

Pander, Christian, 47, 82–83, 97n37

Physiology, 2, 29, 39, 53, 89, 95n22, 150, 156

polygenism, 9, 27n45

Power, 44–47, 53, 56, 59, 61, 63, 78, 82, 85, 90, 100n70, 126, 158, 162–63

Preexistence, 39–42, 45, 67, 71n54

preformation, individual and generic, 13, 47, 64, 100n66, 110, 120n26

Preformationism, 6, 1–13, 38, 42, 58, 68n19, 78, 83–84, 89, 93n13, 97n35

product, 6, 44–45, 102–3, 111

Races, human, 8, 56, 81, 156

reflective judgment, 10, 21, 61–65, 71–73, 85–91, 98n54, 145

regulative / constitutive, 9–12, 15–16, 28n53, 65, 68, 75, 79–80, 88, 125, 150, 154, 157

Rehbock, Philip, 17, 33n92

Reill, Peter Hans, 18, 30n68, 35n103, 66n6, 97n42, 100n70, 165n1

Richards, Robert J., 2, 3, 12, 17–19, 22, 31n70, 33n95, 34n97, 96n30, 132n1, 133n6, 150, 165n1, 169n43

Robinson, Henri Crabb, 153, 166n21

Roe, Shirley, 30n65, 48n3, 66n4, 68n16, 70n40, 93n5, 97n45, 120n26, 132n2

Roger, Jacques, 27n32, 30n45, 93n13, 100n70

Schelling, Friedrich Whilhelm, 16–18, 2, 34n99, 126–30, 132, 146, 150–56

Schiller, Friedrich

Schrader, George, 11, 29n61

Second Analogy, 10, 27

self-organization, 54

sensibility, vital, 5, 58, 93n6, 124–27, 163

Sloan, Phillip R., 2, 13, 17–18, 24n26, 35n30, 27n45, 31n71, 37n100, 57, 63, 68n19, 96n28, 120n26

solidarity, 42, 44, 47

soul, 56, 60, 71n54, 90–91

species, fixity of, 57–58, 61, 64, 112

Spinoza, Benedict, 55, 81, 92, 139–41

spontaneity, 52–54, 59–62

Stahl, Georg Ernest, 11, 14, 46, 83, 90

Steigerwald, John, 98n55

Strawson, Peter, 24n18, 29n58

Systema Naturae, 5, 7, 25n30, 105, 117n18

systematicity, 4–8, 54, 79, 84, 91, 92n1, 94n20

technique of nature, 78, 82, 89, 96n31

teleology, 5, 10, 15–16, 62–63, 88–90, 149–50

teleomechanism, teleomechanistic, 2, 3, 12, 15, 19, 21–22, 37, 123, 125, 132

Theorie der Generation (or *Theoria generationis*), 37–38, 41–44, 75, 82, 88

transcendental philosophy, 5, 8, 12, 14, 27, 58, 60, 65, 126

transformism, 2, 18–19, 35n110, 133n9, 159

transparency, 41, 44, 47

type, 16–19, 45, 81, 130, 142–46, 155–59, 163–64

Vicq d'Azyr, Felix, 21, 129

whole/parts relation, 5–7, 15, 76–77, 85–89, 102, 11, 131, 144

will (intelligent), 55–57

Wolff, Caspar Friedrich, 3, 13–14, 20–21, 37–49, 54, 57, 81–85

Zammito, John, 26n41, 31n66, 36n69, 133n8, 148n20